RELUCTANT HERO

A True Story

by

John Hickman

Order this book online at major online book stores
Or at selected book stores in Australia & New Zealand.
Purchase autographed copies direct from AuthorJohnHickman@gmail.com
This book is featured at www.authorjohnhickman.com

Printed worldwide by Lightning Source

Distribution to the book trade throughout Australia and New Zealand by Dennis Jones & Associates Pty Ltd.

Cover design, metadata, manuscript formatting, templates and photograph enhancement by Publicious Pty Ltd.
www.publicious.com.au

ISBN: 978-0-9870945-1-3 - paperback
ISBN: 978-0-9870945-2-0 - hardback
ISBN: 978-0-9870945-3-7 - ebook

www.authorjohnhickman.com
Published with assistance from InHouse Publishing
www.inhousepublishing.com.au

Dedication

Pilots and aircrew proved to be a rare breed of heroes. Bill's own squadron was wiped out numerically—twice! The average age of fatalities was twenty-two. He knew he'd only the slimmest chance of survival.

When he neared retirement he told me, 'live long and prosper. I'm proud to have come from Notting Hill. It ruined my life, but I'm proud. Prediction is difficult, especially about the future, but had I known then, what I know now, I would never have taken part in World War Two.'

John Hickman 2011

As author it has been my choice to write under my own name. Other names have been changed out of consideration for relatives I have been unable to contact. The writer acknowledges the trademarked status and trademark owners of various products referenced in this work of narrative non-fiction, which have been used without permission.

The publication/use of these trademarks is not authorised, associated with, or sponsored by the trademark owners.

The author states opinions expressed are not his own. Characters, places, brands, media, and incidents are true as communicated to him. To the best of the author's intentions this work is a faithful rendering of the facts as told to him.

As stated, Reluctant Hero is a work of narrative/creative non-fiction. This is a true story as told to me by my family; Mum, Dad, Gran, Gramps, Uncle Charlie, Aunties, Cousins who all knew my dad. Some of Dad's ex RAF crew visited during the 1950s. His bomb-aimer, a sandy haired man gave me an expensive drum set from Harrods department store. When he presented it to me, he laughed and slapped dad's arm. 'Your boy hitting that drum might remind you of knocking noises in engines, Bill.'

I have had remarkable clarity with most of the story. Where memory has failed or a gap has been discovered of which I was unaware, the tone has been recreated with the general content for dramatic flow.

ACKNOWLEDGEMENTS

People who contributed to this story have had to wait a long time to see it appear.

Sadly, for some it has been too long. Many have passed away before publication.

I hope I have done their memories justice.

There are people alive I would like to thank for what has been an historical journey of some magnitude for me to write *RELUCTANT HERO.*

Carole, my wife, is appreciated as always for her patience in tolerating what has become my obsession. She has been a model of calm and reassurance with her recall of our family folklore, as she knew dad for more than thirty years.

Pamela, my sister, the scholar of our family, deserves sincere thanks for her meticulous maintenance over the years of family photographs and records. Our reminiscences were a delightful reflection.

Tricia Eban, my editor, who gave her time unselfishly, kept me focused, and never gave up on me. She prevented my story from becoming a boring diary of events. Her many valuable suggestions and changes allowed Bill's voice to come through me.

Hon. Yvette D'Ath, (Federal Member—Petrie), for valuable guidance through myriads of government departments. Hon Greg Combet and his office (Federal Member—Minister of Defence Personnel) for assistance with government liaison overseas. Air Historical Branch (UK), RAF (UK),

Ministry of Defence (UK), National Archives (UK), Veterans (UK), RAF Disclosures (UK), Service Personnel & Veterans Agency (UK), Ministry of Defence Medal Office (UK), and to those dedicated staff at RAF Cranwell (UK) who prefer not to be named individually but helped me in my labour of love to delve back over sixty years.

Acknowledgement of photographs:
The Photographers who are named; Williams Pioneer Studios Ltd., 529 Oxford St., Marble Arch.
William Frederick Honey (deceased 26/02/1972).
Sara Jayne Fornaro

The other photographs are attributed to unknown photographers.

Other sources:

- The New York Times
- Daily Mirror
- Daily Telegraph
- The Times
- Daily Express
- Wikipedia
- John Hickman 2011

Prologue

This is a story that needs to be told about my dad, Bill.

As a child, I sat mesmerised by what Mum told me about his time as a Lancaster Bomber pilot in the Second World War.

When I grew older, I became interested in the why and the how? Dad was reluctant to talk at first. Perhaps hesitant to give away too many pieces of himself, divulge his inner thoughts, and to relive his involvement in those extraordinary circumstances.

When I asked questions, occasionally he brushed me off, but not too often.

And now back to where my story began, the slums of Notting Hill, London in the 1920s. An area as bleak as the Gorbals in Glasgow. Dad blamed his parents Lily and Fred for his rough childhood. He had issues all right, lots of them, and we're about to go there.

He hated his life, his surname, but most of all his lack of opportunity for a good start in life. Dad became frustrated. He wanted to escape the class-conscious restrictions in which he lived, but didn't know how. He felt inferior, inadequate and these feelings never left him.

Soon a dark side mirrored his rebellious nature. Neither Bill nor his family knew how these traits would influence him in the years ahead although few were not aware the Second World War was on its way.

Neville Chamberlain's efforts and pieces of paper had done little to convince anyone there might be a treaty. When what little goodwill there was ran out, menacing clouds intensified, ready to unleash unspeakable horrors.

War machines started to rumble throughout Europe but no one, least of all Bill, considered what part he might play. A role that would involve frequent death defying missions over hostile Germany, when terror was given a voice that culminated for him, in the awful bombing of *Dresden*.

Chapter 1

BILL'S DARK SIDE 1932 TO 1936

'Get up, Honey, you're a bloody girl,' sneered Alf, as he aimed another kick at Bill's head. But Bill didn't hear his taunt, nor did he feel Alf's boot connect with his jaw. Instead he lay stunned and silent.

It had begun as nothing more than a typical schoolyard confrontation. Three bullies had flushed Bill's conkers down the loo, and then set about him for sport. After fists had come at him out of nowhere, he'd attempted to brush something hard off his cheek. Dazed, he realised, cobbled playgrounds don't move.

Alf prodded Bill with his size twelve boot, 'Come on, Honey. Get up. Fight like a fuckin' man.'

Bill's head was in a whirl as if wrapped in cotton wool. He felt a slow trickle of blood ooze from his mouth. It felt warm and sickly against his tongue.

Where did that come from?

As he lay stunned, a succession of weird notions crossed his mind. All he could see was Alf's boot. He tried to focus. It was dirty and unpolished, battered with torn stitching.

My boots are shinier than yours!

He strained to get a better view of Alf, and when he did, he saw muscles like those of a fighting dog. Tight as cords, flexed in use, as if from his toes up. It was too much for Bill. He lay back with a great sigh and wondered how many men, older and bigger than him, had fallen victim to Alf's kicks. Then frightened of being kicked again, he cowed back defenceless, and shrunk into a ball. Their world blotted out he stared at the school, eyes seared with wanting, hoping a teacher might appear. Why couldn't it be like his favourite fable? In *Alice in Wonderland* the rabbit would appear to save him, but no one came to his aid.

Later at home Bill's mum, Lily, attended his wounds. Quick as her darning needle into one of his trouser patches, she prepared a bowl of warm water from the kettle on the hob, collected cotton wool and Dettol. As she worked her thoughts were of Alf, his crony mates and what they'd done to her only child, 'Shush, I've got you, Bill.'

Lily turned away and bawled for her husband Fred. 'Come look what they've done to your boy.'

Fred was slumped in his chair as if the day had beaten the spirit out of him. He looked up, unsympathetic. He didn't give a damn, 'It's time he learned to stand up for himself, Girl.' He often called Lily, 'Girl.'

'What if he'd died? You'd be sorry then,' cried Lily, embroiled in her own thoughts.

More than Bill's pride took a hit that day, he'd lost four front teeth but his life as a victim continued to deteriorate.

'Honey's a girl! Honey's a girl!' The same chant followed Bill home most afternoons.

When Fred heard about it, he showed no concern. 'He'll have to tough it out, Girl. That or I'll box his ears till he sees stars.'

Lily wiped her hands on her apron. 'You'll do no such thing, Fred.' She smiled a pained expression at the sight of her son's distress, and then swore softly under her breath.

By age ten Bill had few friends at school. Fun was often made of his surname, Honey, which he despised. On those days he hated every one. At home he never felt the support he longed for not even from Lily, but for him as with Fred, there were never extenuating circumstances or shades of grey. Childhood was like that. Bill was punished, then punished again when he complained. Not suited for cheating or deception, he tried not to protest but his rebellious spirit led him off course.

Fred exhaled noisily. 'Falsies are expensive. You'd best learn to bite into your food without them.' To eat an apple Bill quartered it then cut it in to sections. He chewed slowly to include the pips and core and as he did, he inhaled succulent juices he'd not noticed before.

He told Lily. 'They make my nose tingle, Mum.'

'Like a damsel in a rose garden,' laughed Lily. And as a bonus his pocket penknife he carried for the purpose made him feel grown up.

'Trouble is, Mum, without front teeth I can't whistle properly.'

Lily looked thoughtful. 'It's good you can't whistle, Son, because whistling is uncouth.'

Fred put down his newspaper. 'That's right. Whistlers have bad reputations.'

'Why, Dad?'

'Since the times of Dickens, whistlers have been distrusted, Son. It's how pickpockets pass signals to each other.'

That night as Bill lay in his bed, he promised himself when he could afford the best of false teeth he would try to whistle, but then he wrestled with self-doubt. He didn't want people to think poorly of him. What if they thought he was a pickpocket, a villain, a thief?

This led to daydreams about when he might become successful. To be satisfied enough with his station in life to whistle. What in older men might be called silent thoughts?

Next day he asked Fred.

'You shouldn't get ideas above your station in life, Son. Don't grow up wasting your life with pipe-dreams.'

That seemed unfair to Bill. His dad worked as a guard on the railways and was always quick to point out how he'd sorted things out at work, told those bosses a thing or two as without him the trains would never have run on time. Bill wondered, if his dad was only a guard, why was he always full of bullshit?

'I've been on the railways for years. Worked my way up from porter to guard,' boasted Fred.

'You should listen to your dad, Son. He knows best,' sighed Lily.

Bill wondered. Could she be right? He gave her a sideways look, which said eloquently, I don't think so. People are only impressed with him because they're gullible enough to believe anything.

Life was becoming seedier by the day in Notting Hill. The harsh reality was to work hard, live in semi-squalor and get on with it. Most walls carried every howl, sob, scream and crash. Cracks were so big in some places you could talk to your neighbours through them.

'If you're clothed and fed you should accept your lot and think yerself lucky. And yer find a job—any job,' ranted Fred.

Since 1932 the gathering storm of military matters had preoccupied the nation. Many were aware the country was rearming, about to prepare for war.

'Those bloody Germans are at it again. I tell yer the only good German is a dead German.'

'If it's not the Germans, it'd be those wretched French,' sniffed Lily.

December saw an unprecedented 1,250,000 copies of the Daily Mirror sold with news of the appointment of British army chiefs and threats from Japan. Ten editions were distributed in London alone. Fred brought a newspaper home. 'It's not looking good, Girl.'

The British government supported the notion of how the Japanese air force comprised antiquated aeroplanes flown by pilots with poor eyesight.

'Their rice diet prevents them from flying above 5,000 feet,' quipped Fred.

But the commonsense of these statements was never questioned.

Lily's older brother, Charlie, was different from the rest of their family. A well-to-do bookie, he lived in an upmarket semi-detached in leafy West Kensington. Drove a brand-new Ford saloon car that cost £100, smoked big cigars and looked bastard-arse rich. Bill admired him because he was special and different. He wore bow ties, striped shirts, had a loud voice and acted really smart.

Charlie enjoyed The Daily Telegraph crossword, which he'd submitted for years. There were good prizes to the value of two guineas and Uncle Charlie had won twice. Admiration for him shone in Lily's eyes whenever his name was mentioned.

Deep down Bill nursed a secret wish his own dad could be more like Uncle Charlie, especially as he gave generous gifts at Christmas.

December 1933 saw Bill score a brand new cricket bat from Uncle Charlie. He'd never been close to a proper one, or watched the game played, but he'd seen newsreels.

Fred told him, 'It's a game rich people play on a village green. Pretty to watch and they dress in matching whites.'

When Bill held that bat he felt different, as if transported to another world. He forgot his name was Honey with tatty hand-me-downs. Socks with so many darns he was embarrassed to remove his shoes. Patches in his backside that made sitting down uncomfortable. How Bill longed to be on a village green dressed in resplendent white. And when he gripped his bat he was. Not stuck in Notting Hill.

He played out his fantasy in an unhurried and gentlemanlike manner. In his mind spectators lounged in deck chairs and watched him as he took his place at the crease. On an enormous clubhouse veranda people enjoyed afternoon tea and when Bill scored his first century they all stood as one, and applauded him.

Bill snapped out of his daydream, back to reality. Those gentry weren't on their knees crawling between horses' shuffling feet, dodging being kicked. Yesterday, he'd chased after a ball in a filthy cobbled street. Dodging horses' hooves as they backed away was one thing, sliding about in their piddle and poo, quite another.

'Mind you don't fall under a wagon, Bill,' warned Lily.

'There's not much chance of that, Girl. He's quick when he wants to be, as fast as if the devil's after him.'

As days passed into weeks Bill became so enthralled with his bat, it turned into something of a status symbol. He carried it with him everywhere. Like a holy grail of relics he knew it could never be replaced.

'You certainly know how to stand out from the crowd, Son. Why don't you leave that silly bat at home?' But Bill didn't listen to his mum. Why should he? When he carried his bat he lived the dream.

Alf and his bullies were a quarrelsome bunch. More used to being sent to bed with a mouthful of knuckles from their drunken fathers than fantasising about cricket. When they saw Bill with his bat tucked under his arm they picked on him. Perhaps to them the bat represented a sign of betterment contrary to their surroundings.

That or they were bored shitless and didn't like cricket.

'Come here, yer sweet little toothless bastard. We'll have that fuckin' bat, sweet Willy Honey. It's too fuckin' good for the likes of you,' sneered Alf.

Bill knew Alf was a nasty item. He had cruel features, accentuated by baring his yellow or blackened teeth as he laughed. Bill thought his face looked like a vulture's with blackheads and more zits than anyone else, because he never washed properly.

Before Bill knew what had happened, Alf and his mates had him helpless. He couldn't move, his arms and wrists were pinned firm. Stan, another shitty piece of work and Alf's best mate, goaded Bill then aimed a blow at his head. His swing missed but Bill caught his foul breath full on. He felt ill.

'I'd rather have no teeth at all than rotten ones like yours,' taunted Bill.

Stan tightened his grip. He disliked Bill for more reasons than the bat. At home he rarely had anything more than a piece of toast for his dinner, let alone tripe and offal, as he knew Bill did. Added to that, word on the street was when Stan's father wasn't drunk he was dead drunk, whereas Bill's father Fred was known to be sober. Unbeknown to Bill, in Stan's nightmares his family were only one step from the poor house.

Jack, who had small crooked features, the type that looked as if they'd been added as an afterthought, held Bill from behind. An arm of his short strong body tightened around Bill's neck, which meant now Bill could hardly breathe.

Déjà vu. He felt as helpless as he had over his conkers. Victim of their sour breath and loud voices, they laughed out loud as Alf wrenched Bill's prized bat from his hand smiling a gap-toothed leer.

'We'll have that, yer little honey sucking shit!'

'No, gimme me back me bat,' yelled Bill in panic. 'It's mine! Me Uncle Charlie gave it me for Christmas.'

'Oh, did he then, we'll see about that.' Alf weighed the bat in his hand. His cold glance skewered at Bill. 'It's a nice bat, sweet honey boy.'

Alf was big for his age, a giant of a boy without a neck. His arms had the girth of other men's legs; his hands the size of dinner plates were adorned with large callused knuckles and sausage-like-fingers.

'I'll tell me Dad on you,' whined Bill.

'Tell who yer like, sweet Willy boy. We're gonna make you very fuckin' unhappy.'

'I'm already fuckin' unhappy,' snapped Bill, as he tried to wrestle free.

Bill watched powerless as Alf stepped up to the old iron railings that separated the basements from the pavement. They stood like sentinels in a long straight line only with breaks for entry gates and steps. He heard Jack's hard bark of laughter, then silence.

Alf re-weighed Bill's bat in his hand as he eyed the railings and as he did he took pleasure at Bill's tormented face.

'No!' shouted Bill.

Alf brought Bill's bat down hard on the first railing and waited to see Bill's pained expression. Content with the result he slowly, deliberately continued to hammer Bill's bat along the top of each iron railing.

Bill winced as each strike tore at the unseasoned surface of willow.

Alf slowed, he enjoyed Bill's distress but by the fifth strike Bill was sobbing with anger and frustration, forced to watch helplessly while his tormentor pummelled his prized possession.

'You fuckers!' shouted Bill. 'You'd better have eyes in the back of your heads. I'll get yer for this. You're bigger than me, Alf and I know I can't beat yer fair, but I'll get back at yer somehow!'

Alf stopped dead in his tracks. His face paled at such an unexpected reaction from Bill. Stan and Jack were uneasy too. Stan rotated on the spot, his mouth open in astonishment. Bill's eyes were wild, his face flushed. He shook from head to foot. Shaken, they released Bill and backed off.

'Here, take your fuckin' bat then, little Willy Honeykins,' Alf sneered, as he threw it down at Bill's feet. Something about Bill's reaction, his viciousness towards them had unsettled big Alf and his bully mates.

'It's only a fuckin' bat,' muttered Stan.

Bill looked at them each in turn. He knew this moment would be etched forever in his mind. 'Yeah, but it's my bat. You had no right!'

'Yeah, well yer can have your bat, or what's left of it. We've got better things to do,' said Alf.

The gang sauntered off, whistling and cat calling to their mates. Bill was left standing alone in the street, his breathing ragged. He made a pugnacious fist but no one cared.

'I told you to put that silly thing down and leave it here,' scolded Lily, when Bill arrived home. 'If you'd done as you're told, none of this would have happened. You'd better not let Uncle Charlie see what you've done to his bat.'

'It's not his bat, Mum. It's mine. He gave it me. I didn't start it, Mum. I couldn't help it. They're bigger than me.'

Fred's eyebrows bristled like a dog's hackles. 'Be quiet. Stop arguing with your mum. You're giving my arse headache,'

Over the top of his newspaper he stared at Bill. He saw a runt of a lad, not unlike how he remembered himself at that age. Fred and his father before him had grown up in the slums. He'd known nothing else. He'd never been a fighter himself, but he knew to shy away from a fight was unthinkable. As he pondered Bill's situation, he felt bad for his son.

'Surely, no father wants to see his son worse off than himself,' goaded Lily.

Fred put down his paper and gazed into the hearth for a while, almost as if he was searching out some hidden message in the flames.

'When your brother gave Bill that bat he gave him an activity. Instead he's carried the bloody thing around with him like a disease. But all said and done, Alf's done more than steal someone's spoons this time, Girl.'

In that moment Fred had decided to advise Bill how to defend himself.

'Look upon this more as a tactical retreat, than defeat, Son.'

Tears of anger and frustration welled into Bill's eyes.

'Never go looking for a fight. That's wrong, but if one comes looking for you that's different. You must win by fair or foul. You hear me?'

'But they're bigger than me, Dad.'

'Forget Queensbury rules. Size doesn't matter. Think foul, fouler the better. Strike a decisive blow because if yer don't you'll end up the loser.

Believe me, the bigger they are, the harder they fall. At the end of the day, size has nothing to do with it.'

'But how, Dad?'

'Pick-up whatever's to hand, Son. Better make sure it's something heavy and start swinging.'

'Whatever are you teaching him, Fred? I've never heard the like,' scowled Lily.

Fred ignored her. 'Always go for their ringleader, Son. He's easy to spot.'

'Biggest mouth and trousers, you mean,' Lily muttered from deep in her washing up.

'Bring him down, any way you can, Son. Eyeball him and keep the fucker down. You end it, Son. And fast. If nothing else, kick him hard in his balls. That'll fix him.'

'Fred. Watch your language in the house! You know I don't approve.'

'Sorry, Girl. But the boy needs help.'

'And not before time. Him almost without a tooth in his head. Judging from the size of that Alf there might be value in your Dad's advice. That or an approach from behind.'

'Your mum's right, Son. But sometimes I wonder what I've married into.'

Lily shook her head and wiped her hands on her apron. 'I know little of those things, but that's how Charlie protects his patch at the track.'

Fred turned back to the hob, pulled his chair closer to the blaze. Lily joined him and in silence as sweet as music they had a cup of tea.

Lily frowned. 'That Alf's a danger to the public.'

'So is the dentist, Girl, but he roams free.'

Bill brooded for days before his darker side emerged. There was almost a constraint in the air when he took his bat out from under his bed and held it firm. As he swung it back and forth angry tears of white-hot rage welled up inside him. He swore retribution would be his. There were no shades of grey only black and white.

Not much of a cricket bat anymore, but it's a fine weapon, he thought.

Bill knew he needed to find Alf and bring him down.

'Bigger they are, the harder they fall,' Fred had said.

That night Bill hardly slept but when he did, he dreamed of victory over Alf and his bullying mates. Towards dawn he dozed to the muted sigh of the wind in their chimney.

When he rose he shook from head to foot, but not from the cold. It was anticipation and excitement at what lay ahead of him. Now it made perfect sense. Fred was right. If he could defeat Alf, their ringleader, he shouldn't have to fight Stan or Jack. Bill was determined to get those three monkeys off his back once and for all. The question was no longer, if he could defeat Alf, but when. And the when, he'd decided, was now.

His bat held firmly under his arm he hoped for the best but planned for the worst.

At least I can't lose any more front teeth.

Driven by ferocious determination, Bill took long, slow breaths, and then began to search the streets for Alf, his tormentor.

Chapter 2

BILL'S REVENGE SUMMER 1936

Bill spotted Alf rolling a smoke outside the school toilets. Engrossed with chatting up a girl, he never noticed when Bill stepped quietly around the back. As he stood silent in the shadows, Bill plotted his attack. Strong toilet smells and coping with his sweaty fear began to turn his stomach.

Not a good time to retch.

His wait seemed eternal. He clutched his bat until his grip was so firm he had trouble loosening it. Finally the girl became bored, wandered off and left Alf alone with his smoke. The coast was clear. Bill knew he would never have a better moment to seize his opportunity but fear of failure almost consumed him. If Alf sensed a problem or pre-empted Bill's attack, he knew he would lose. The bigger boy would disarm Bill and thrash him severely.

Bill took one last deep breath. No turning back now! He stepped out of cover with his raised bat and lashed out at the big boy's knee.

Alf saw the bat coming but distracted with his smoke, too late. Whack! The bat connected. Alf let out a shriek and went down like a sack of spuds. That's easy, thought Bill. *He's down! Now to keep him down.*

'You little fucker!' cried Alf.

Bill didn't hesitate. His dad's words rang in his ears. 'You end it, Son. And fast.'

He knew to win he had to finish it. He brought his bat down so hard on Alf's foot he thought he heard it crush his toes. Alf groaned and stayed down with a whimper. Bill hit him again hard across the fleshy part of his back. He heard—a dull splat, like a slab of raw meat dropped onto the counter at the butcher's shop. He realised he dare not get too carried away. If he hit Alf on the head he might kill him and although he wanted to, he was afraid of the consequences.

Down came Bill's bat, again and again. Woofs and sighs as he pummelled into Alf. Every sinew in Bill's body was pumping. Every now and again, a wet clap followed by a groan.

'Try and get up,' roared Bill. 'Come on, fight like a fuckin' man!' But Alf couldn't. He started to sob. His eyes went bloodshot.

'Please, Bill. Stop! You're hurting me.'

'At least yer called me Bill. I'm not sweet little Willy Honeykins any more, am I? You big useless fucker!'

Alf inclined his head. His mouth gaped open in astonishment at the punishment he'd received.

Bill paused, rotated on the spot. His eyes wild. 'Your reputation's bad, Alf. If yer don't want to defend yourself like a man, here's one more. Arsehole!'

This time Bill swung his bat higher and harder. Thump! The big one! He hit Alf across his right hand. Alf screamed with pain.

Bill smelled victory and it was sweet. He promised more and worse to come with his clenched fist. 'If yer ever come near me again, I swear, I'll fuckin' kill yer, Alf. Yer hear that?'

There was a long dazed silence. Bill felt a muffled nothingness, as though he had watched this scene played out from inside a glass bubble. Suddenly he felt exhausted.

Alf knew he was in trouble. The second day would be worse and nothing could stop bruises from showing the third day. Alf sucked in his breath. He looked pitiful. 'I'm sorry about your bat, Bill. I'll never bloody do that again'

'Do yer promise?'

'I promise!'

Bill felt sudden remorse but tried not to show it. Alf's tear stained face and lank brown hair no longer made him look threatening. Bill straightened up and as he did Alf flinched and cowered back.

Bill placed his bat back under his arm. 'You and I are through, Alf. Unless you or your bully mates want to make it otherwise.'

Bill didn't walk away; he swaggered, pugnacious and arrogant with victory.

Next day Alf was in awful shape. He felt so crushed and disjointed, he wondered what joints might be broken. He had more discolouration than if he'd been wrapped in cases of fruit fallen off a lorry. His knee cap swelled-up and swirling purple galaxies of blue-black bruises spread across his entire back, which he kept hidden. His damaged hand was twice its normal size and throbbed like hell. As he limped about his breathing was ragged; word was he passed blood in his urine for days.

When he was asked what had happened to him, Alf evoked the Notting Hill code of silence. No one suspected Bill. Everyone thought Alf's comeuppance was from bigger boys he'd upset. After much shaking of heads the matter quickly became old news.

From then on Bill changed. He became more confident and strutted around like a cock in a henhouse but always with his comforter, the bat. When Alf's crew were nearby he turned and faced them. Bat high in the air ready to strike without taking his eyes off theirs. But as predicted, Alf and his mates gave Bill a wide berth and left him alone.

'For Christ's sake leave him be,' Alf told his cronies. 'If yer don't, he'll come up behind yer with that bloody bat in his hand.'

'He's not that big. We could always take his bat from him, again,' suggested Stan.

'Fuck no! Without his bat he might use something worse,' said Jack. 'Like a claw hammer.'

'Just leave him alone,' groaned Alf.

The tables had turned and Bill savoured his hold over them. Their talk near Bill was conducted in knowing nods and shrugs and whenever they could they avoided him.

Bill became cocky and baited them.

'Come on then you arseholes. Let's see how tough you are now. Weekend warriors are yer? I dare yer. Try to take me bat and I'll fuckin' kill yer with it!'

'Fuck-off, Billy boy,' Stan said nervously, 'we've got no fight with you.'

Bill continued to provoke them. 'Love me or hate me, yer bastards. But yer can't ignore me! If yer do, I'll have yer.'

And Bill, without any provocation, would wade into them with his bat, scattering them and causing those big bully-boys to back-off and run away.

Like kicked mongrel dogs they looked for easier prey to intimidate. They never messed with Bill again.

Lily was supportive. 'Whoever Alf's God is in his mercy, He might forgive him for your missing teeth,' she sniffed. 'But I never will.'

'Thing is, Dad, when the worst of the damage was being done to him, it didn't sound that bad.'

'I told you, Son. Sometimes it's more blessed to give than receive,' smirked Fred.

Chapter 3

CUT-THROAT 1937 TO 1939

Bill did what was expected of him and left school at fourteen.

'It's all to do with combinations of seven, Son,' Fred beamed. 'At seven you're a child. Fourteen is twice seven. Time to move on. Most men marry and have kids by twenty-one. That's three times seven. You're back's buggered by fifty-six, that's eight times seven. And you'll be pushing up daisies by seventy.'

'Ten times seven,' laughed Bill.

Bill craved to be successful but didn't know how. He wanted to belong, to fit in, but for him that became a worry. In his topsy-turvy way, he felt sure to remain in Notting Hill, meant he had to stagnate. And that he refused to do. His emerging competitive nature became stonewalled. He felt lost, up against formidable odds, while his parents worked as hard as ever to pay rent and provide food on their table.

'What I can't stomach, Mum, is hypocrisy, idleness or fraud,' raged Bill.

'No, Son. You tell them!' You're dumber than a bag of wet mice, she thought.

But Bill had decided he didn't want to be part of life in Notting Hill. His problem was how to achieve it. If only it could be as easy as when *Alice in Wonderland* followed the rabbit. He was about to exercise an exceptional will to improve and get out of the slums, to find something better.

'It's not only the greyness and the drizzle of the weather depresses me, Dad.'

'You can't escape that, Son. Not even your Uncle Charlie does that by living in a posh suburb.'

'It's the greyness of the people, Dad. They look unwashed.'

'Working class coarseness, yer mean. We're not good enough for yer, Son?'

'It's not that. What's happened to all the greenery and trees we only seem to see in picture books? Why in the parks are we forbidden to walk on the grass?'

A lack of answers and the deafening sound of silence, threatened to capsize Bill.

'I'm dragging myself up by my bootstraps,' announced Bill. That was how he liked to explain his slow progress. It was an exceptional attitude for a man so young.

He begged Uncle Charlie for an opportunity in his betting business but was turned down flat. 'It's a risky business, Bill. Up one day, down the next. I'll be getting out soon myself, with a bit of luck before I lose everything. I'm doing you a favour, Son, I really am, by not taking you in.'

Bill was devastated. He'd thought fish might fly before his Uncle Charlie turned against him. A major escape door had closed in his face. Rejection by his uncle was a new experience for Bill and the hurt cut him deeply. * 1st Footnote

Short on qualifications Bill decided to become an apprentice of something. He chose a manufacturing jeweller in Portobello Road. They used concepts of basic arithmetic. Unlikely to induce any awful recurring headaches, thought Bill.

'Bill's got it made, Girl.'

They were sure he had, until a critical day when Bill used cyanide to clean gold unsupervised. Without thought he placed his tobacco pipe down on the workbench. Its stem, the part he put into his mouth, had come into contact with the smallest trace of deadly poison and he almost died.

Bill came to, dizzy and disorientated. His mind clouded with half-forgotten names of long dead relatives. He'd been laid-out on the floor amid spinning rooms and nausea for several minutes. When he recovered, he was wiser and none the worse for the experience but hungry to learn.

There were clouded hints of self-improvement without support of his under-arm symbol—the bat. He continued to educate himself. The flame to learn burned brightly but only one flaw could be improved at a time. The opportunity of going back to full-time school was rarer than wedding tackle on an action man.

At sixteen Bill had health issues; awful abdominal pains. Doctors misdiagnosed he'd poisoned himself at work then realised he had appendicitis. An appendix operation was major surgery in 1939. Bill was lucky to survive.

After convalescence he reviewed his dental plan. He acquired false upper and lower dentures. It was a big improvement at meal times but when he whistled, it was never quite the same tune.

New teeth or not he felt cursed with a natural pessimism, which he blamed on being born in the slums. This became his Achilles heel. Disillusioned by what he perceived as dread came upon him in waves, similar to grief. He feared mediocrity, which threatened to overwhelm him. Scared of sliding into obscurity in Notting Hill, Bill couldn't wait to plan his great escape. But how could he make it happen? As he pondered he became more aware of his every failure. He saw clear contrasts between the brightness of his dreams and the utter botch-up of their carry through. But this was to become an even greater recipe for deep despair.

At home Lily and Fred never talked about what interested Bill. He became a loner. He took long walks unaccompanied and became an avid reader. Technology, high-class fashion magazines, diets and clothes fascinated Bill. That better educated people tried to lose weight intrigued him. His mum Lily could have benefited from losing a few stone although Bill never said anything to offend her. She was a rotund jolly woman with enormous breasts and a continuous sniff.

As a kid she'd pull Bill towards her to give him a big, big hug and he disappeared laughing in to the warmth and security of the giant folds of her bosom. A beautiful woman inside and out, all four foot nothing of her. Her warm green eyes danced with humour and when she let her hair down it was so long she could sit on it. But years of living in Notting Hill had taken their toll. Lily's hair had turned prematurely grey and worn up in a bun, made her figure resemble a massive upside down light bulb. Lily often poked fun at her own appearance. 'My problem was being born with duck's disease.'

Another feature was her feet being too small, so tiny she often toppled over but when she did she blamed her bunions. Her enormous breasts might have contributed, but she never mentioned them. Her falls became legendary during Bill's growing years but were never serious. No broken bones; only grazed dimpled knees, which were invisible to everyone except Fred.

Bill thought Lily could have given some of her bulk to Fred and then some, but neither Lily nor Fred ever pursued good health. Nobody did in Notting Hill. No one ever went to a gym unless they were athletic like boxers in a ring. And no one ever jogged or ran anywhere, not unless they were late for work or had a bus or train to catch. Lily walked to their local shops or caught a number 52 bus near the end of their street.

Fred cycled to work on an ordinary black-framed bike with a big basket on the front without a chain guard. He wore cycle leg clips to keep his uniform trousers from flapping into the well-greased chain. When he returned home Fred hoist his cross-bar up onto his shoulder and carried his bike of about eighty pounds weight up six flights of stairs to their little flat on the third floor.

'That and a day's work is exercise enough for me today,' quipped Fred.

He stood his bike against a wall, often in their bedroom. If left downstairs in the passageway it would have been easy pickings from the street.

Fred smoked un-tipped Senior Service cigarettes when he could afford to buy them. Otherwise he rolled his own. Most people smoked but Lily didn't. They were ordinary people Lily and Fred, and they drank and ate whatever they could afford.

Bill's attempts at improvement were ridiculed whenever someone found out. To him those bloody goal posts were being moved only to make life more difficult for him. Dimly aware whenever he entered a room he made little more impression than a draft he soon became ultra depressed and moody.

Bill turned against the Church. 'History is scattered with so called holy men who lived like lords, fat on the offerings of poor people. Through ignorance they bought blessings rather than bread for their children. That's not true Christian conduct.'

Fred agreed with him. 'If there is a God, He's got a lot to answer for. I for one would like to sit in on His judgment, Girl.'

'For hundreds of years the Church has held the pen,' ranted Bill. 'And what they wrote is supposed to be fact. I believe it's mostly lies, written by dishonest men who suppressed the history they didn't like and wrote what was more favourable to them.'

In September 1939 Britain declared war on Germany and along came the Second World War. Over one million men would serve in Germany's air defences, with another million and a half providing ground defences.

Uncle Charlie shook his head in disbelief. 'Goes to prove, if there is a God, He must have a weird sense of humour, I suppose.'

'Personal tragedy has no time-table,' sniffed Lily. 'I wonder whatever will become of us all?'

Fred lit another cigarette. 'It's been proven throughout history, it's not a real war unless someone's making a good profit. I'll bet this one will turn out to be a real doozy.'

Chapter 4

DECISION TIME 1940 TO 1941

In February 1940 Bill was seventeen years of age and a touch less than six feet tall. His body was not yet filled out to fit his frame.

'He's hollow in the chest, not much weight behind him, Girl.'

'Skinny as spaghetti, you mean. Lean as, with that raw-boned look of an adolescent.'

But despite his physique, Bill was about to enter the most extraordinary period of his life and prove to be a man useful to employ.

Britain braced for German invasion. March 1940 to September 1941 the Luftwaffe bombed targets in Britain at will. From as far apart as London to Liverpool, Glasgow to Portsmouth they reigned supreme. The primary aim of most British plans during the summer of 1940 became defensive. London and several other major cities came under attack from enemy bombers. More than 40,000 British civilians were killed in London alone during the Blitz. An average toll of two hundred and fifty fatalities were recorded a day following hits over sixty consecutive nights.

'Our troops abroad are getting about as much support from home as measured by a fart in a thunderstorm,' ranted Fred. Lily made a face.

In May 1941, Egyptian born Nazi Rudolf Hess, Hitler's deputy and third in command, flew to Britain on a peace mission. The *Daily Telegraph* reported he baled out over Scotland at 6,000 feet and told a Scottish farmer in impeccable English, 'I have an important message from Hitler for the Duke of Hamilton.'

Accused of displaying signs of mental disability he was swiftly declared insane and locked away in solitary confinement as reward for his odyssey.

'Something's not quite right, Girl. It smells suspicious to me.' * 2nd Footnote

Bill's eighteenth birthday approached. With ears like radar dishes he sat spellbound listening to Cousin Eric, on leave from the infantry. 'Conditions at the front line are bloody awful enough to frighten anyone.'

Fred shook his head and lit a smoke. 'Despite propaganda bullshit, news from there is never good two days in a row.'

Eric went on to recount his horrors of the trenches, the mud, drenching rain and death in stupefying detail.

Lily bent from the waist to pick up her sewing; the exertion seemed to leave her breathless. Her lips retreated to thin lines. 'It's enough to swear anyone off having babies. They're cannon-fodder for the British government, that's all they are.'

Then Eric started to lose it. 'You've got no idea what it's like. It's bloody terrible.' His torment came in waves. Lily cried as he talked.

When he finished, there was a long pause.

'I know it's hard out there, Eric. I knew those trenches in the last war,' Fred said quietly.

But nothing deterred Eric. 'We worry we'll be killed in a bayonet charge and all the while we lay in mud. We're lucky to get any food at all. When we do everything tastes the same. Yuk! Even our smokes get wet.'

At home Eric spent most of his time in idleness. He cried often. When he visited Lily and Fred he warmed himself at their hob and stared unseeing at the floor. Occasionally he gazed out of a window. Otherwise he fought to hold back his sobs, which upset Lily big time, and embarrassed Fred. All this scared the crap out of Bill. As to what lay in front of Eric, his demons were beyond anyone else's control. He was too upset and frightened.

'I don't want to go back, Lily. It's terrible. I don't want to die,' sobbed Eric.

'But you have to go back, Eric. You can't change your mind. It's too late now.'

Fred steered Lily out of the room. 'Not much I can say to the kid, Girl. He's barely in control of his own bladder.'

Lily was in tears. She composed herself and they rejoined Eric.

'It's not as if you picked a bad day for it. If you don't return to active duty and they catch you, they'll shoot you for sure,' Fred said.

Now Eric understood. At fourteen, he'd lied about his age, as many had and joined up. Enthralled with excitement and filled with a sense of adventure he'd seen war as a free ticket to see Europe and get laid.

'You had no idea what you were getting yourself into,' said Lily.

Fred changed tact. 'What people don't understand about being in the army; is its inaction. When men aren't scavenging for food or smokes, they're in trenches full of water. All because some idiot educated beyond his intelligence said they should be there.'

'You're right, Fred,' said Eric. He lightened up. 'We were shifted in the middle of the night once by another nincompoop. Corporal said it was a less than defensive position from where we'd been.'

'We never slept properly, our equipment was always defective,' continued Fred. 'And when supplies arrived, if they ever did, they were never what we expected.'

Bill was candid. 'Now I know why I'm terrified of being drafted into the army.'

Fred frowned. 'History will confirm those in charge have been lax with birth dates. But mark my words sending children as young as fourteen to the front line won't be talked about much in the future. They'll keep that under wraps for sure.'

'Poor babies,' sniffed Lily. 'They say it's being called to arms. It's their mother's arms they're wanting.'

Now Eric had had a taste of reality, he wished he'd been less hasty.

Lily tried to mother him.

'There,' she said, 'I've made you a nice pack of your favourite sandwiches.'

Fred rolled him some smokes and gave him pocket money from their rent jar on the mantelpiece. They both tried to cheer him up. Fred told him one of his favourite stories from his war, the First World War.

'It was the war intended to end all wars, Eric. At least that's what they told us then. Back when radio communication was in its infancy, it was hard to understand. According to the crackle, 'Reinforcements are coming and we're going to advance.' Or was that, 'Three and four pence and we're going to a dance?'

Everyone laughed, even Eric.

When they saw him off from the railway station the next day, the three men shook hands and Lily kissed him on the cheek. 'Be strong. You'll be fine, Eric,' she whispered.

They waved and promised to love each other forever but Lily couldn't hold back her tears. * 3rd Footnote

After he'd gone, Bill was unable to shake memories of Eric from his mind.

Fred held court. 'It might be simpler if we stayed in England and shot 5,000 armed men every week, Girl.'

Bill's morbid sense of foreboding had him worried. He visualised desperate unwashed men at the front, sweating with anger and fear as he had with Alf. He racked his brains for a less physical alternative but time was against him. He didn't want to wait to be called up for fear his freedom of choice would be taken from him. If he was conscripted into the Army he knew he would be up against Germany's finest and he knew Eric's prospects of survival were poor.

'Trouble is those bloody Germans are all hulks of troopers, Son.'

Death had never been busier and would become more so, but thoughts of being skewered by an eight-foot Hun death machine in hand-to-hand combat, with or without his trusty cricket bat, caused Bill to wake up most nights in a cold sweat.

'You're behaving worse than a whore in church,' laughed Fred.

Bill's recurring nightmare was being effortlessly attached to the pointy end of a German bayonet and flicked around for amusement as if he were a giant Catherine wheel.

'I wouldn't be surprised,' said Lily, 'If that big boy Alf doesn't join the Army. He likes to fight and they'd be a profit in it for him.'

'Judging by the size of him,' laughed Fred. 'They should have a war, just to employ him.'

'Worse in the Navy, Son,' frowned Lily. 'There you'd be trapped on the rolling deep.'

Bill imagined visions of him being permanently seasick. While a delightful addition to the crews' endless amusement, this clouded Bill's judgment.

'Of course in the Navy, you'll still die, Son,' sniffed Lily.

'But instead of playing ring-a-ring-a-roses with German bayonets those seas will rise up to swallow me whole,' frowned Bill.

'No mercy out there, Son,' Fred shook his head. 'Not with the briny less than friendly.'

'They say drowning is a pleasant death, although I can't see it myself,' sniffed Lily.

'Bloody hell no, Girl. The cold, the exhaustion, they can keep that to themselves. I'd rather take a bullet.'

Cold sweats and nightmares continued to exhaust Bill.

If he wasn't dodging giant Hun troopers he was desperate to keep his head above the icy cold wobbly stuff.

Father and son sat quietly by the hob after Lily had taken herself off to bed. For a while they tried not to disturb the air they breathed.

'Fuck those bayonets, Dad. They leave a hole!'

'Army's out then, Son.'

'And a pox on their damn ships!'

'Nothing nautical then, Son.'

The British Government promised everything possible to attract young men to arms. Bright new buttons on their uniforms, even a free haircut. Things were hotting up. There was an opportunity to trade an empty stomach and tatty clothes for a sense of belonging.

Death and misadventure were not promoted.

'They make going to war look romantic,' sniffed Lily.

'Provided they overlook dying for their meal ticket, Girl.'

In London it was as if everyone rallied to the flag.

'There's more flags than at a dawn service,' quipped Fred.

Men in uniform were everywhere and the ladies were easily impressed, even by boys as fashionable as Sabre Toothed Tigers in flared trousers.

The propaganda machine was in full flight. Bands marched, wirelesses blared. Who wouldn't want to take part? Bill began to relax.

Fred shook his head. 'Those bloody politicians are on their band wagon and everyone's being primed to fight.'

'The kids you mean—not the politicians,' corrected Lily.

Who wouldn't want to make that ultimate sacrifice? These were crazy times.

After filling adventurous days with killing there was an added perk. Volunteers would be de-mobbed immediately following the cessation of hostilities. This guarantee of an early discharge was seen as a massive advantage not to be missed.

'First out of the services will beat the long queues for the best jobs,' Fred said.

Bill thought he'd be spoilt for job offers. He perked up. But he was to learn the hard way, how politicians are for the most part a waste of skin. This pledge, as with many others, would never be kept. Assurances were only given to achieve their own short-term agendas.

Advertisements portrayed aircrew as supremely glamorous. *Odol Toothpaste* showed a delighted RAF officer with a pretty woman on his arm and urged: *Cheer up and smile—keep smiling even if you're running risks.*

What risks? Their testosterone fortress only had to defeat Hitler's hairy little one-balled arse, which shouldn't take too long, His Majesty promised.

The Battle of Britain was around the corner, as was the Luftwaffe under Field Marshall Hermann Goering, an ace fighter pilot from the First

World War. Weighted down by his Iron Cross for bravery but now a little too large in the bum to squeeze into a cockpit, he burned with speeches. Mainly about how he would throw more planes and bombs at England than anyone had ever done before him. And in that he would succeed.

Germans were a formidable foe. Their armaments were supreme, their discipline extreme. They were superior Teutonic beings who did everything better than everyone else.

The Luftwaffe had no such concept as a tour. German pilots didn't retire after thirty missions or ops (short for operations as the RAF called them.) They flew until they died or survived the war but German morale was higher than high. Their military intelligence ran circles around the British, as did their technology.

Hitler was on a roll. He promised to wring England's neck like that of a chicken.

'Some chicken; some neck,' quoted Winston Churchill.

'Churchill's up for the top job, and he does nothing for anyone except on cash terms,' said Fred.

'He's good at speeches,' retorted Lily. 'I'll give him that. But I'm not sure dying is preferable to kowtowing to those Germans.'

'No! Nor drive their wretched Volkswagens, or own a Dachshund, Girl.'

* * * * *

...We shall go to the end. We shall fight in France, we shall fight in the seas and oceans, we shall fight with growing confidence and growing strength in the air, we shall defend our island whatever the cost may be. We shall fight on the beaches, we shall fight on the landing grounds, we shall fight in the fields and in the streets, we shall fight in the hills; we shall never surrender...

Winston Churchill, 4 June 1940

(An excerpt from his speech in the House of Commons, after he became Prime Minister).

* * * * *

It was on for one and all as paranoia gathered momentum in triumphant procession.

'Wherever Churchill goes people flock to cheer him,' Lily said.

'People,' sneered Fred. 'We'll cheer anyone, if frightened enough. And don't forget Churchill needs to break the old guard, Girl. New brooms replace the old.'

Rationing was in and the first items under the hammer were bacon, eggs, tinned ham, butter, sugar and precious icing sugar. To begin with the Germans had an underdeveloped air force but by late summer 1941 they'd seriously upgraded.

Nazi Germany dominated continental Europe. August 1941 saw over one hundred British aircraft lost. The following month was worse with losses of over one hundred and thirty. Many were shot down over English cities. One in four British planes perished but worse was to come. October saw more massive losses.

The Germans missed their East End docks target. Instead their bombs dropped on London. Hitler became angry that civilians had been attacked unnecessarily and issued an apology to the British Government.

Did Churchill see this as an opportunity to accelerate hostilities in his favour?

And was that why he ordered the RAF to attack Berlin as a reprisal?

An angry Hitler then ordered the Blitz on London.

Oh, well. Nobody's perfect, thought Bill.

Often on smoggy days in London it was impossible to see a few yards ahead. Other times a heavy overcast sky appeared laden with snow, when it wasn't. Even on those better days distance was relative as any horizon became lost in a sea of smog. Pilots reported visibility reduced to less than a few hundred yards. The days grew shorter as winter approached. Children went to school and came home in the dark. Any warming sun might appear sporadically about noon and disappear soon afterwards.

It was on a miserable day like that in January, with dismal thoughts about the weather in mind; Bill came to terms with reality. In the face of so much horror, he'd decided not to lie about his age as Eric had done.

One month before his eighteenth birthday at Lord's Cricket Ground in London, Bill joined with hundreds of other would be hopefuls, for duty at the feet of a bunch of aeroplane enthusiasts. His Majesty's King George VI's Royal Air Force.

Bill filled out their masses of forms and stood in line for selection. He knew the next few hours might well be a turning point in his life. Aware time, once behind him, could never be retrieved or revisited his mood became mixed. He reasoned as a Leading Aircraftman he should be far removed from strapping eight-foot Hun troopers with pointy bits. If by some stroke of luck he made the elevated status of flight crew he would be even further removed. Bill had lifted his hands for emphasis. 'Way, way up higher than high. And far, far above any rolling deep seas too.'

Bill completely forgot his fear of heights. At home he was incapable of standing on a chair to change a light bulb without assistance. But if his decision needed any reinforcement, there was an additional carrot. Payment for qualified pilots was a dizzy nineteen shillings and sixpence—per day! To Bill that was a worthy pinnacle.

Good money for the boy from the slums especially with more than a million people unemployed, thought Bill.

There were medicals to be conducted and then issue of an enormous pile of kit.

Like other local lads Bill was to be billeted at home as accommodations were reserved for those from out of town. Some luxurious north London flats were used, which belonged to exceedingly rich Jewish princesses. Bill looked on with envy. If only he'd been an out of town lad. Luxurious before they were stripped of everything flash, as in fancy fittings.

Even carpets were lifted before the rabble moved in. Colonial and country lads alike sat in bathroom sanctuaries second to none and used unlimited hot water from taps not blanked off—yet.

Bill's anticipation and excitement continued to mount. Before him, he knew; would be an extreme test, both professionally and socially. The higher they raised the bar, the closer to the ground he felt. The boy from the slums was having troubles but he wasn't the only one.

Many quietly suspected this wasn't going to be a walk in the park. Nor would it be an opportunity to sit around thinking about sex every day. But any such talk might smack of negativity, even treason.

In the First World War many good brave men were executed for far less by firing squad. They were made an example of to ensure the rank and file toed the line. A timely reminder of how any discussions about what might go wrong with the strategies of the powers-that-be, be kept to whispers and shared in private with only family and close friends who could be trusted.

Bill was barely eighteen years old, when full of beans but clueless, he looked about him in the examination room. He felt about as useless as a screw-in light bulb for a bayonet fixture. What to do? He might as well be attached to the sharp end of a Hun trooper's bayonet. Most of those who competed against him for prospective aircrew positions were from better backgrounds. They could hardly have come from worse. Some had entire houses to themselves with lots of chimneys, even their own bathroom with hot and cold running water. The majority were from middle class families. Above them, aristocratic elite, who wore better uniforms with officer status and Oxbridge accents.

Back at home Bill said to his dad. 'Makes them sound as if they're talking with a plum stuck in their mouth. They're so bloody wah, wah; their talk is so affected it's hard to understand what they're saying.'

'Few of those upper classes will see much combat, Son. As with the world over you'll find wealthy kids are rarely drafted. They're easily identified as better educated when they speak.'

But everyone was better educated than Bill. Some were on leave from universities others had fancy double-barrelled names heralding aristocratic and privileged origins.

Bill felt leaden footed and totally inadequate. For him there was the measured collapse of one expectation after another. He wrestled with his doubts. Why on earth would the Air Force ever choose him over this lot? Bill's ambition to become aircrew, perhaps a pilot had been formulated over cups of hot sweet tea in his parents' scullery kitchen. Tea stewed the colour of his pipe tobacco from a teapot rarely allowed to run dry, two sizes bigger than his head. And all in a small flat, with only one smoking chimney. While Lily had ironed and folded the washing, Fred had washed and shaved over their kitchen sink. Bill chatted excitedly about his plans how he might become a fighter pilot to avoid death in the trenches.

Fred was enthusiastic. 'Sounds a good idea, Son. 'If yer can pull it off.'

It had been a big day for smiles and nods, but what had seemed logical then, had fast become foolhardy, now a distant mirage.

Bill rubbed his sweating palms on his knees and tried to concentrate. Pondering his fate with the exams, he twiddled with his pencil. He felt stonewalled again. Perhaps he should give up.

CHAPTER 5

WEEDING-OUT

Bill broke from his muse of negativity. With the world locked in mortal combat he shouldn't expect life to be easy. But if only this exam was as effortless as when he defeated Alf.

He'd been so nervous about call-up he'd enrolled in evening classes where he achieved more than expected. Gone to the top of the class and allowed to give out the pencils. He'd felt prepared to face anything—until now.

Might his competitive edge be about to fail him?

Being government inspired tests, the majority of questions bordered on absurd. They looked baffling. There was one in particular he read quietly aloud, again:

'A rectangular diagram of a field with shaded blobs to represent trees accompanied this question. Underground warrens were shown with dotted lines. Dimensions, compass points and entry for said rabbit was clearly marked with a big X.'

Confused and in a state of rising panic Bill looked around the room. This test was insurmountable but everyone else appeared busy and confident. Some were diligently making frenzied calculations, others engaged in Pythagoras theories. A few longhaired intellectuals had slide-rules.

Bill's discontent only helped him remember why he should have worn clean underwear.

His idea of *pi* despite evening classes was something to have with chips and peas. He'd always counted on his fingers and still did, content to pull down A's and B's in sincerity and honesty. He cast his mind back to questions at his evening classes.

One question had been; *how far could a train travel into a tunnel?*

It was a trick question, thrown into the mix as light relief and not to be taken seriously.

The answer they wanted was; *half-way* (because after the mid-way point the train would be on its way out).

Surely not, thought Bill. But the question was clear; 'How far can a rabbit run into a field?'

Was it possible the only answer the RAF wanted was same as with the train?

He took a chance and wrote *half-way.*

The next paper they handed him was even more confusing. Large letters at question number one stated. 'FIRST READ ALL the questions slowly and thoroughly.

Do not attempt to start this test until you have read the final question.'

Bill was impatient. He went straight to the final question number thirty.

It stated. 'You are not required to answer any questions in this test. Please place your name here and sit back with your arms folded. You have completed the test.'

As he looked about him, a few sat bolt upright their arms folded and smiling.

Others were writing like mad.

As Bill bumbled along he started to feel as if fate had rolled out a welcome mat.

The kid from the slums might scrape through their tests without too many trips to the toilet.

When he found out he'd passed their Mathematics Exam and General Knowledge test, his excitement flooded over.

At home Fred was surprised. 'I didn't think yer had it in yer, Son. Well done.'

For the first time in Bill's short life he felt a sense of accomplishment. *But wait on.* Why would they choose him over bigger, stronger, better-educated and more athletic looking applicants?

'I don't know whether to laugh, cry, or go jump in the fuckin' Thames, Dad.'

More surprises were to come.

He needn't have worried as the bravado types were quickly eliminated soon after the rabbit had run into the field. Like rabbits running out again, they were assigned to other duties and trades, including His Majesty's kitchens. Big strong men convinced they would be taught to fly were trained to wash dishes or peel potatoes the RAF way. Their dreams of soaring above the clouds evaporated with the steam as they washed plates.

Whatever strategies the RAF and its powers-that-be implemented in their selection criteria, the gung-ho types were allocated elsewhere.

Some candidates appeared to be straight from Central Casting. One swaggered around Base with a revolver strapped to his hip, with an ego to match. Another wore cowboy boots stretched almost to his scrotum. Against them Bill felt inferior.

It's a certainty they'll get tickets to fly before me.

On Sunday 7 December 1941 the Japanese had attacked Pearl Harbour and the USA officially entered the Second World War.

'This place is half as big as New York Cemetery, Bud, and twice as dead,' laughed Cowboy Joe. His handshake crushed Bill's hand until his toes curled up. 'Our President Roosevelt will soon sort these fuckin' Nazis out!'

Worldly, strong and clever, thought Bill. And he wished he were more like Cowboy Joe. Some had panache for the aeronautical, sounded confident and knowledgeable about machines that flew. How he wished he were more like them. Others had experience, if only as a passenger and wore outrageous flying apparel. Another had done eight hours solo.

Some sported bushy style moustaches waxed to perfection. If they didn't resemble Clark Gable, they made Dick Tracy look like a pansy. They knew everybody worth knowing. One even spoke of a get together while skiing in St Moritz. They all had an edge over Bill. They walked the walk, looked the look and spoke the part.

A couple sported suave leather jackets like rebels without a cause. He envied them their longer hair and rebellious clothes, which he couldn't afford and wasn't allowed to have. There was even a robust Australian proud of his tough guy tribal tattoos splattered over his arms.

When Bill told Fred, he laughed. 'So tough, you mean he has them emblazoned on his testicles.'

'Ouch! Enough to make your hair curl and arse whistle, Dad,'

Bill reasoned this put them at the front of his queue. All were certain to be accepted. How could these men, who looked and sounded so much better, be overlooked in favour of him? His depression set in again with a vengeance. This had to be a waste of time. Why had he been so foolish as to think he would be considered for the ultimate privilege to fly?

Some thirty would-be aviators were left in his group. And of course one of them felt he should make conversation. Bill didn't want to chat. He was so nervous he needed to pee about every ten minutes. *Why is there always one bull shitter in any group?*

The chap continued to prattle on non-stop despite no one listening to anything he had to say. So much so, Bill began to feel sorry for him.

Then, one-by-one each of these assured, knowledgeable and robust champions, many who perhaps had more guts than common sense were weeded out for supper duty.

Soon the only candidates who remained were the pimply faced, round-shouldered, unfit looking specimens with Bill ably leading this questionable group.

Each and every one of them had one thing in common. They were cowards.

No typical gung-ho types remained. No smart-arse bravado types at all. Certainly there were no heroes. Cowboy Joe was gone. *His President Roosevelt hadn't helped him.*

And with him, tattoo Aussie. No testosterone filled, gold medal gods remained. Only a few frightened young men, who for their individual reasons, had decided they wanted to fly. From this un-inspirational looking group, final selections for pilot training were made and to every-one's amazement, most specimens squeezed through.

One common denominator remained. They were seriously afraid.

Bill asked an officer why?

'They were eliminated because they were fearless or stupid. True courage is to be scared stiff, but keep going anyway,' he sneered.

Bill had never considered this aspect although he realised fear drove his feet.

'So, to control fear means you're truly brave.'

'You're smarter than you look, which probably isn't a good thing,' said the Officer. 'Anyone frightened shitless, who performs his duty is a brave man indeed.'

I've got the frightened shitless part by the balls. All I need is the follow through.

Bill felt sure he now had a chance. He'd scraped through their exams and survived their weird discriminating processes. He was in. The boy from the slums of Notting Hill had beaten the odds.

Tuesday 20 January 1942 was another milestone for Bill. He was selected for pilot training thirteen days before his nineteenth birthday.

March 1942 saw the first raid by the new Avro Lancaster over Essen in the central Ruhr Valley but for Bill, next came the breaking-in weeks.

'I want three volunteers who can play a piano,' announced the Corporal.

Absolute silence.

'Come on, lads. Surely some of you are musical.'

Not an eyelid flickered. Then a young man weakened. 'I can play the accordion, Corp.'

'Well done, lad. Another two.'

'I've had lessons, Corp,' said another. 'I can play a tune on a recorder,' offered a third.

'That's capital, lads, capital. Now I've my three, and I never had to resort to nominating anyone.'

His eyes glistened with anticipation. 'Get your kit up off the floor and follow me. There's furniture to move. That includes a piano.'

Boot camp had the usual basic disciplines, drill and constant yelling of orders by corporals. There was a lot of shouting and marching in the RAF. Shouters had smiles like brass plates on coffins. Every now and again, Bill saw a Sergeant.

One in particular was Sergeant Matthews. He was an unsmiling disciplinarian of colossal innate charisma, all gaunt six feet six inches of him. A caricature for who discipline came before talent. His thickset shoulders under his craggy countenance augmented Bill's opinion of him. Corporals dealt with minor transgressions but watch out when Sergeant Matthews became involved. A far more daunting experience than addressing smarmy officers in their handmade uniforms as 'Sir.'

Marching, physical training, and lots more marching with a high measure of verbal abuse in the primeval form of control was doled out to excess. Those who shouted had extraordinarily powerful voices. For sheer volume they're bloody impressive, thought Bill.

The marching felt endless. In between were lectures and physical training, then more marching and more physical training. Shoes had to be shined until they glowed but on the upside; meals were plentiful. Another downside was inoculations.

Bill winced and turned away as the needle penetrated his skin. 'I won't look, Doctor.'

'Only one of us has to,' said the doctor.

Young men who watched invariably ended up unwell. They toppled from their chairs to lie in crumpled heaps on the floor, amid the merriment of others.

At the commencement of a typical day's march, everyone groaned.

'Come on, lads. Not far today,' chirped the corporal.

No one believed him anymore. He'd said that every day for a week.

Bill badly needed a rest and a smoke, as did Oliver.

'Aye, my name is Ol-iv-er,' he beamed. 'But as I want to spare you effort for marching, Bill. You can call me Olly.'

They laughed.

Olly was a clear skinned ruddy-faced young man from the Yorkshire Dales. He had cheerful brown eyes that danced when he spoke and carefully groomed red hair.

His principal problem was his rosy complexion, which fought hard against creases of a tan. 'Aye, I'm not built for warmer weather, Bill.'

Olly had not been exposed much to fresh air. At Glasgow Veterinary College his days were spent locked away in a dilapidated three-storey building, once a pumping station. There, within the confines of their less than hallowed walls, he'd been too busy studying to venture out.

He looked dependable enough. Pity his Yorkshire Dale's brogue was contaminated with Scottish. Otherwise when he spoke Bill conjured pictures as stodgy and unromantic as Lily's puddings. Usually Olly had a lazy grin that seesawed across his face, but not today.

'Aye, I hope it's naught much further, Bill,' grumbled Olly. 'We're nowhere near end of day and already I'm history.'

'I heard that, lad,' shouted the corporal. 'Come on, lads. Lift 'em up. Faster. At the double, now- one, two, one, two. And once more we go around the park. Let's go! Lift them feet. You'll thank me for this later on.'

'It's hard to appreciate his style of kindness,' gasped Bill. He understood their passion for shrieking abuse as that ensured no one dozed off. What he couldn't understand was why, if he was going to fly a plane, the need for so much marching.

'Thank Christ we're not foot soldiers. There might be an element of fitness involved, but we won't be carrying the damn aeroplane on our back.'

'Aye, this running is a sure waste of energy, Bill. I've never heard of anyone outrun summat like a bullet.'

Bill's new boots gave him hell. His feet suffered and he developed athlete's foot but he was so proud of his new boots and socks he refused to part from them.

'Best cure ever for sore feet, Son, is to pee on them,' Fred had said.

Bill turned his nose up.

'Well, it is only your pee, Son, it's not as if it's someone else's. And it's clean when it leaves your body.'

Olly agreed their daily marching was relentless.

'I've been marching so long, my arse has changed shape.'

But none of this really bothered Bill. He was as fit as a butcher's dog and twice as determined. He couldn't care less. He was in and for now, he wasn't simply on cloud nine—he owned it.

Christmas was harsh. No presents from Uncle Charlie and not much cheer on anyone's tables. It was worse overseas; Norwegian tables were bare even of fish, as the Germans had taken their herring catch. French tables and cupboards were pitifully dressed, as the conquerors had depleted their food supplies. Greece, Poland, Holland and Belgium cities were populated by melancholy, hollow cheeked women and children. Unbeknown to Bill, the American war machine was about to shift into high gear.

CHAPTER 6

PRIVACY SEPTEMBER 1942.

Initial Training Wing and Elementary Flying School

But it wasn't all roses. Bill felt his competitive spirit had helped him through, but dropping a third testicle would have been preferable to communing with nature the RAF way. On a more than daily basis he suffered their open-plan toilets.

'Even in the slums we had a closable door, Olly. Granted it was sometimes minimal, like string looped over a nail, but at least it closed.'

Olly was unconcerned but without a barrier between his new world and himself, Bill felt defenceless and unhappy.

'Sitting on the crapper in full view, I might as well be squat on a bog in the middle of bloody Oxford Street. It's the final straw.'

'Aye, but never mind, it's naught for long.'

'Long! There I am about to punch out the poo of my life while the entire bloody world saunters past.'

Olly changed the subject. 'Aye, but those doing the shouting are passionate about their jobs. They're never out of naught but yelling range for long.'

'Yes. But they're qualified to shit in private!'

'Aye, they're naught but bastards.'

'Nothing a few minutes alone with me and my cricket bat wouldn't fix.'

'Aye, they should be in party hats for sure. For them it's summat of a big round of fun and revelry, by gum.'

'For everyone else it's a pity party if you ask me, Olly.'

'Aye, at least until someone loses an eye.'

Much as the Drill Sergeant would have liked to fit a small portion of his over polished boot up each cadet's backside, his military masters didn't allow him to do so.

Damn those witnesses, thought Bill.

To touch was against the rules. Instead a lot of ducking, weaving and over enthusiastic throwing was learned well. The powers encouraged corporals to kick large empty metal dustbins between bunks at reveille. Or hose icy cold water over someone to make a point of control, but no actual touching.

This style of fun and revelry was unpopular with the college boy fraternity but not much of a problem for Bill, except having to crap in public.

At home to wash Bill added hot water from their kettle on the hob to cold in a bowl. Their only scullery tap was cold water, and they shared a privy outside in the back yard. Now Bill had two taps; hot and cold.

Once a corporal stood too close to Bill's bed at 0600 when he bellowed in his recumbent ear. 'Let's be having you, you mother's little sleeping beauty, you. Rise and shine.'

Awoken from a deep sleep, to the accompaniment of crashing dustbins and shock of an unprepared sprinkle of icy cold water, Bill felt like he was coming out of a coma. He panicked and lashed out. He caught the corporal a blow to his nose and he bled like a stuck pig.

Bill apologised profusely. He'd meant no harm but on his next patrol the corporal employed a six-foot buffer zone.

As a result of this incident it was ruled when awoken by His Majesty's footmen, men asleep couldn't and shouldn't be accountable for their actions. Bill was not punished for his attack on the corporal.

Olly opened up to Bill like a steamed mussel. He elevated him to a status Bill didn't believe he deserved. Between them they formed the opinion most of those engaged in training were content to sit the war out in comparative safety.

'The old adage of those who can—do and those who can't—teach comes to mind,' quoted Bill.

'Aye, stubbing their toes on an errant dustbin, hoarse from shouting too much, or getting a chill around too much cold water comprises naught but their highest risk of injury,' laughed Olly.

Bill and Olly agreed; the louder and more boisterous the pricks were, the less likely they would ever see combat. In this and perhaps and for the shortest time, Bill was in tune with the philosophy of His Majesty's Royal Air Force. Sadly he was unaware of it.

Cadets were taught to handle a variety of firearms. This included how to load, safe handling, actual firing, and how to dismantle and clean.

Bill's favourite was the Sten submachine gun as there was little to dismantle or clean. What looked like a piece of errant water pipe had a bolt with a sort of a bent fold away handle and a bullet clip.

'It must be the worst weapon ever conceived, Olly.'

'Aye, might be cheap to manufacture but it's bloody unreliable.'

Not surprisingly, because it was easy to clean, it became Bill's favourite gun.

He became proud and confident and was able to dismantle and re-build in one easy action. As the gun was poorly designed it remained jammed most of the time. As a result it was rarely discharged.

'Proper cleaning is hardly ever necessary, Olly.'

'Aye, Bill, but when it does fire, it's summat like your own personal cloud of lead.'

'I see no reason for pilots to be skilled with hand guns, Olly.'

'Aye. We're in the RAF. If we get that close to enemy, it's summat of a problem.'

From a wet and muddy trench, Bill was instructed to put down his Sten submachine gun and throw a live hand grenade.

With pin pulled, it slipped from his less than stable fingers. On its altered trajectory said grenade narrowly missed the corporal's head, the very same who had copped a reveille blow from Bill in their dormitory.

There was an awful fuss. The corporal wanted to prepare a noose with Bill's name attached and enlist help to flog Bill within an inch of his life.

But being mindful of military rules, he was forbidden by his masters to do so. Instead he took a biased dislike to Bill.

'Aye, naught but a marvel of ecstatic hatred, he's got for you, Bill.'

'I can see he's thinking, and I know exactly what,' laughed Bill.

'Aye, he's behaving like summat's stuck up his nostril.'

And so, for reasons not clear to Bill, his comrades in arms elevated him to dizzy heights. Olly led the contingent and stirred the group in Bill's honour. The more Bill protested his innocence and lack of premeditated thought the more they praised him. Without effort Bill became a status symbol nicknamed Ballsy Bill—with a grenade.

Bill's new don't-you-mess-with-me attitude rested well with him, until Sergeant Matthews sent for him.

'You the Cadet who beats up on my corporal?'

'No, Sergeant.'

Papers on the big man's desk scuffed as he swept his arm sideways.

'You saying, Cadet, you didn't throw a hand grenade at him?'

'Well, yes, Sir. I did throw it. I intended to throw it, Sir. That's what I was told to do,' said Bill defensively. 'I couldn't help it, Sir. It slipped out of my hand.'

Under cross- examination Bill admitted he didn't like the corporal.

'No one likes corporals, Cadet. That's why they're corporals. But we don't want to kill any, do we? The Germans are trying to do that.'

Bill was horrified. 'Yes. But Sergeant, I never meant to injure him, let alone kill him.'

A few decent grovels and the slightest hint of distress worked wonders.

'There's nothing I can do to you, Cadet the Germans won't do better.

Behave yourself. I don't want to hear another single peep about you. You got that?'

'Yes, Sergeant.'

'Otherwise I'll have your arse nailed to the floorboards. Now, fuck off! You're dismissed.'

'Yes. Sergeant. Thank you, Sergeant.'

The case against Bill never eventuated.

Merriment continued without respite but there were threats made by corporals.

'If you don't end up peeling potatoes in the Navy, you'll be taught how to kill by numbers. Got it?'

Bill was reminded of Alf and his cronies. But corporals had a licence to bully.

'You're a horrible little man. You couldn't kill a German if your life depended on it,' said the corporal.

After Bill allowed himself to be summarily broken, he learned to be happy the RAF way.

He had clean sheets, new blue uniforms and a pay slip of less than £1 a week.

Food was plentiful three times a day and always the flesh of some sort of animal even if covered with gritty gravy like Thames mud. Bill reasoned this was not a bad way to live except for his butt nakedness in public toilets.

'They could also lighten up on their obstacle course, Olly. Fred says he never saw a single twenty foot wall with rope hung over the side of it. Not in his war.'

But Bill's stint near home base neared a close. Excitement was in the air. Changes were about to happen. News of his posting seemed too good to be true. He was assigned to another Training Wing but news of where, was slow to come through.

'Aye, Bill, I'm to be admitted to same Wing, by gum.'

'The air is thick with rumours, Olly.'

True. First they were off to Aberystwyth in Wales. Then it was to Newquay in Cornwall. Bill understood general strategies of the powers were unlikely to take his preferences into consideration but in the final analysis he hoped he might not be too far from home. Already he missed it.

Chapter 7

BILL'S JOURNEY MAY 1943

When the news broke, it was not the slag heaps of Wales, nor the pirates of Penzance, Cornwall. They were off to Canada.

'Bloody hell! Where's Canada?' asked Olly.

'I'm not sure, Olly.'

Neither Bill nor Olly wanted prolonged farewells, which was fortunate as no one came to see them off.

Ensconced into the deep bowels of a munitions ship bound for Canada, they left Southampton 29 May 1943. Lily was proud of her son. He was going abroad. Even her well-to-do brother, Charlie had never been overseas. No one in the family had been beyond Southend-On-Sea for holidays.

Olly braced himself for the sea air. 'Aye, it'll be a welcome change to go for a cruise.'

As the deck rails crammed with teary-eyed teenagers Bill and Olly explored below.

A blast from their ship's horn signalled departure. Soon after they left the quay, the wind stiffened. In their wake seagulls circled and cried out. Bill thought they sounded like lost souls.

Not long and it was Bill's cue to take his own special place at the deck rail. There he looked windswept and according to Olly, smelled interesting.

'Aye, I'll keep you company, Bill. But I'll stay naught but upwind.'

Their sea passage was rough. As Bill leaned ever further to seaward he reminded himself between heaves. 'I joined—the Air Force—to avoid—all things—fuckin' nautical.'

Their ship suffered not only from a seesaw motion but a yaw. When they went below to eat or sleep, Olly was unaffected and ate as if he was a junkyard dog.

'Does nothing put you off food?' asked Bill. He looked green.

But Olly's eyes were already drawn to the cheese plate.

'Don't you ever stop, Olly? That cheese is pitted and wobbles, bit like our faces after a big night out.'

'Everyone's in constant fear of being torpedoed, Bill. We might as well eat.'

Wrapped in his lifejacket, Bill braced against the ship's motion while he pined for his white cliffs.

'Aye, some are so bloody scared they never take off their life-jackets, Bill.'

But Bill was impervious to Olly's taunts.

Olly refused on principle to wear his lifejacket. 'There's naught point. If summat happens and ship goes down out here, we might as well go down with her.'

Others smoked, read and wrote letters home while between times they turned green too but Olly was an exception. He was never seasick as he had a strong stomach.

While at the ship's rail Bill realised no one wanted to chat. Others moved away rather than be drawn into Olly's conversation.

'Aye, if fear of being torpedoed isn't bad enough there's worry of being trapped behind them huge waterproof doors.'

'They're designed to close fast if water rushes in, Olly. Trouble is,' continued Bill, between heaves. 'No-one knows for sure, which side—of the doors—will be safe?'

'Aye, everywhere below smells fetid and closed up.'

'That's because—it is—Olly!' Bill composed himself. 'Hell, none of this was on their brochure.'

The sky became blue as they approached the mighty continent. After days that felt like weeks they disembarked. Bill's seasickness lifted the moment his feet touched shore.

'Never was I so pleased to get off that fuckin' ship, Olly. It was a joyless trip all round.'

'Aye, but did you not offer up prayers while on board, Bill?' smirked Olly.

'I'll offer up prayers to anyone until I'm on dry land, Olly.'

After an extended train journey they arrived at Saskatoon in Saskatchewan. Slap bang in the middle of wheat country.

The winter cold took on a new meaning whereas in summer the mercury soared to levels unheard of at home. Bill and Olly settled with no trouble into their new domestic groove.

On a Military Base run in cooperation by Allied forces they adjusted to this land of climatic extremes.

'Aye, it's the kind of cold freezes your snot and makes your bits and pieces shrivel up and disappear,' said Olly.

In summer everything shone so much sunspots pulsed red and black across their field of vision. Mind-melting heat screwed with their brains. That and anything in between was the norm in Saskatoon. To the boy from Notting Hill, the sheer size of everything about Canada took his breath away and Saskatonians were friendly.

Near Base was a small town.

'Aye, or at least a place with shops and a hospital,' laughed Olly.

They saw a big sign, Drink Canada Dry.

'Aye, Bill, let's give it our best shot.'

Bill liked Canadians but some thought English people were uptight.

'Aye, they're not wrong,' agreed Olly.

Canada was light years ahead of what Bill had left behind in London. Instead of narrow cobbled streets there were wide straight tarmac roads with modern motorcars instead of decrepit, horses and carts. Without horse dung, pavements shone like marble.

From Base to town was a short bus ride and en route stood a lonely wheat silo.

Olly craned his neck to see the top of it. 'Aye, it's so tall it reaches almost to the heavens.'

Silhouetted against the clear blue sky it stood out as a beacon and dominated the skyline.

'You're right, Olly. It towers above the other buildings. I've never seen anything like it. I don't think we have those in England.'

Then as the silo loomed closer, and closer, the town itself came into perspective. Beside a turn off to this one and only grain storage was a large sign with an arrow pointed to the obvious—To the Silo.

Olly and Bill thought that was hilarious.

'Aye, it's not likely to go to any other place, is it?' laughed Olly.

On the silo in large letters was painted the abbreviated name of their town S'toon.

'Not a bad idea if someone's lost,' commented Bill.

After what seemed an eternity in the heat of the crowded bus. 'Everybody out! It's the end of the line. This is as far as we go,' bellowed the driver.

An unfair description, thought Bill. End of the line, as he put it, didn't look so bad.

Although compared to London, Olly was right. It was a small place with shops.

Saskatoon wasn't a town of blue bloods, but had a fair share of blue-collar backgrounds that strived towards middle-class.

At Base their three square meals a day changed from the reliable meat and three vegetables to better cuts; superb roasts and exotic fish like salmon.

Olly could hardly contain his excitement. 'I naught miss their gritty Thames gravy. That fish was superb. Aye, look I've almost finished it.'

'If I were a religious man, I'd pray for a miracle,' quipped Bill.

'Don't blaspheme, might prove unlucky.'

'All right then, I'll have apple pie and ice cream instead.'

At home Bill had lived on a cheap diet of potatoes, stodgy puddings, tripe and offal as in hearts, liver and kidneys. He preferred the softer square cut from the middle of the Yorkshire puddings but Lily's rule had always been whoever ate most pudding could have most meat. For Fred a meal without meat never quite hit the spot. Lily knew those who ate most squares of pudding, with a generous serve of gravy, would have less appetite for the expensive meat. But there had always been plenty of fresh greens on their plate, especially cheap cow cabbage.

In their Mess a steward asked Bill, 'Would you like fresh flapjacks, Sir?'

Comfort food. Good stuff. Bring it on. And he liked being called Sir.

All he did was nod. He aroused interest when he ate his apple with his penknife.

'Naught seen anyone eat an apple like that before,' laughed Olly.

A magazine called Blighty was popular for drawings of pretty girls. It carried a cartoon of a cinema usherette asking a woman on the arm of a pilot saying, 'I haven't got two seats together, but would the lady care for a seat next to another airman?'

For Bill the sun shone on high every day. It was cloudless without a sign of greyness or drizzle. Framed in its endless crystal, clear blue, birds cart wheeled about.

Much as I hope to be doing soon, thought Bill.

It became the norm to imbibe and slurp copious quantities of alcohol, mainly beer, whisky, gin, even vodka.

Olly enjoyed a vodka. 'Aye, it's naught but a Russian holiday, Bill.'

Bill developed a taste for whisky with soda.

'It tastes bloody good, Olly. Even from a tin cup.'

They acquired the art of crunching in a classy manner on a potato chip.

Bill savoured new tastes and had no trouble convincing himself he was doing his bit to support the local economy. He was introduced to the novelty of tea bags and blueberry and apple pie, helped down with lashings of extra creamy ice cream.

For a short time his thoughts ranged back to Eric. What a shame he hadn't been assigned to this. Certainly here was nothing to cry about.

In winter it was so cold the indicator arms on cars froze. Before a right turn, the driver hurriedly opened and closed his door. To turn left, they did. Unlike England, houses were built almost entirely of timber but better designed, well insulated and double-glazed. Inside the front door was an entry vestibule where you changed clothes coming in or out, without affecting living conditions.

Canadian homes were snug in winter and cool in summer, unlike the slums of Notting Hill. Back home in the flat Bill often turned sideways to let someone pass due to lack of space. Here there was a feeling of vastness.

Central heating was new to Bill. Efficient use of cheap coal dust was transported from a storage area, with what looked like a giant worm screwed into the furnace. No heavy coal scuttles to fill and cart from basement coalholes to upstairs. And there was less mess. Heat was distributed through radiators to maintain an even temperature.

Life seemed better organised than what he was used to and provided luck went his way, he felt sure he would soon learn to soar like an eagle.

Bill gambled at cards and tried his hand at golf. Not long and he'd reaffirmed he was useless at the one and unlucky at the other. 'It's an expensive game of marbles if ever I've seen one. Destined to ruin any man's self esteem, Olly.'

Olly didn't agree. He enjoyed a game of golf.

Bill had never been in favour of tattoos and thanked his lucky stars he'd never succumbed to those dubious badges, as in 'L.O.V.E.' or 'H.A.T.E.' on his knuckles. No tribal tattoos or licence numbers would be attached to any part of his anatomy. Nor was 'left leg' going to be affixed comically to his right.

Another cadet, Richard, had applied for a short-term commission. A bit of a toff, he appeared to move in a perfumed cloud. Trim from all angles he'd been sent to RAF Desford, England. He was reassigned to Canada when the powers realised home was too dangerous for aristocrats, due to an overactive Luftwaffe.

Richard agreed with Bill. 'Ignoring significant problems like being captured and wanting to be someone else, the question remains, why? From a purely practical point of view, old chap, why would anyone want to stamp themselves with such horrible candle-snuffing words mindful of some indelible idea they once had?'

Bill was super impressed with the way Richard expressed himself. 'You know, you're right. Makes perfect sense.'

'Doesn't help much if you need to disappear or change your identity,' winked Richard.

Olly nodded in agreement. 'Aye, and a bloody good argument against getting summat like a tattoo in first place, if you ask me.'

All three comrades-in-arms agreed, and took an oath of allegiance then and there. None of them would get drunk and add any form of captions to their skins. And should one of them waver the other two would be self-sacrificing and rescue him from his insanity.

Bill started to feel comfortable. He'd never enjoyed any degree of camaraderie. In Olly and Richard's company, for the first time in his life, he felt enthusiastic to participate.

Richard was everything Bill imagined the epitome of an ideal Oxford University graduate to be. Pompous, yes, but he was not the usual chinless wonder. He fitted Bill's mould of how the perfect head master should behave in an expensive school. When they first shook hands it had been firm but friendly. He'd looked Bill squarely in the eye, and Bill liked that too. The sort of gaze that belonged on an older man but almost as if he had a light around him. Bill understood why, with his privileged background, the RAF had earmarked him for a commission. He carried himself so well it was unthinkable he didn't already have a batman to look after him. Heterosexual, privileged and intelligent, thought Bill.

Bill wondered at the absurdity of him and Olly trying to rescue Richard from having a tattoo, if he wanted one. Quiet and well spoken he was without doubt the most English sounding man Bill had ever heard.

'Aye, sounds perfect fuckin' Oxbridge to me, old chap,' mimicked Olly.

'Richard behaves more as if he's on a sabbatical than about to go to war.'

In a way they all were.

'In hindsight, chaps, perhaps I was ill advised to interrupt my promising career as a professor of philosophy to volunteer and pursue this dream of madness,' announced Richard thoughtfully. 'I mean. Do I really want to join the RAF and become a pilot? Then again, it would be nice to learn to fly at their expense.'

Olly suspected another reason Richard might have been keen to leave England was the break from other men's wives, who from oblique mention in discussions, pursued him.

'Aye, Bill, we're in the presence of a true walking penis. He's summat of a working theory, for sure.'

Admiration for their well-groomed home grown aristocrat with the large bushy style droopy moustache shone in their eyes. There was something special about Richard. Without doubt he was a ladies' man. From his groomed hair down to his trimmed fingernails he displayed impeccable manners. Whenever he came into social interaction with women even if he had no intention of asking them out, he was a traditionalist. Old-school rules applied. He held doors open for busy cooks, moved errant chairs out of the way for waitresses and always maintained a degree of sociable eye contact. There was always a clutch of adoring ladies who lived to be in his presence.

In this way he occupied himself with women in much the same way as an angler might toy with a fish. Sometimes he little by little reeled them in; otherwise he'd let them run a while. Any who might not enjoy his sport could break away. Like any good sports fisherman he revelled in using light line for his purpose that could be easily broken-off. In contrast to his aristocratic ways, he avoided whenever possible wearing a tie or buttoning his shirt at the neck, around which hung a thick silver chain. From it swung a heavy medallion in the shape of a large key.

'Aye, what's the fuckin' big key for, Richard?' asked Olly.

'A trinket, old chap, given me by Pater to commemorate my twenty-first.'

Sometimes Richard's fingers rested on the key, and as with any habit he would fondle or pull at it. That he came from money was obvious but he was a socialist at heart.

An unusual combination, thought Bill. A true noblesse oblige who cared for underdogs, and was more supportive of Bill's struggle with his lower working class background, than most. Bill felt whenever he came into contact with Richard he was in the presence of a quicker and livelier mind than his own. How gifted, and to have such fine prospects at his feet.

Bill wished he was more like Richard who became something of an unofficial mentor to him. He confided with Richard about his feelings towards his dad, Fred. Richard seemed to understand.

'Your dislike for your father is unceasing and all consuming, Bill. But only because your world has been too small, my friend, and within, your father assumes an outsized significance.'

How profound and clever, to be able to put the sentiment into those fine words.

Over the next few weeks it was as if they were back at school. They were given homework and plenty of it. Richard took it all in his stride but Olly and Bill struggled.

'Did you do any revision for your tests, Olly?' asked an instructor kindly.

'Aye, some, Sir.' Olly looked worried.

'Congratulations! You hid it well.' Their instructor shook his head sadly.

Others were less congenial and took every opportunity to point out their shortcomings, which annoyed Olly and Bill.

'A fool shows his annoyance at once, but a prudent man overlooks an insult,' whispered Richard.

Over a few drinks in their Mess the three comrades came to know each other better.

'Aye, I went to Glasgow Veterinary College because it was only place would take me. Their finances about as shaky as their academic standards.'

'I was at Oxford University when I wrote to the Air Ministry,' recalled Richard. 'I told them I would take leave of absence if I could fly an aeroplane. It took about six months to get a reply with an application form. Their letter was signed, 'Your most obedient servant.' I thought at the time, what delightful manners the Royal Air Force have.'

Olly and Richard laughed but Bill looked serious. 'I left school at fourteen without any qualifications to do anything. You could say due to my deficient scholastic abilities and obvious lack of toast master skills in Notting Hill, I became an apprentice manufacturing jeweller—to eat.'

They nodded.

'Is Notting Hill a place?' asked Richard.

'Yes. But if you found yourself in Notting Hill, you wouldn't like it there. It's as if even the sun doesn't hang around unless it has to.'

It was apparent the three colleagues were from entirely different backgrounds. Bill might have come from the poorest part of London but now he lived and worked in the company of others from better origins. He could see an opportunity to make friends on a level par, as their blue uniforms of the same rank bore no distinction. But inside Bill felt more as if he stood out like a prostitute among brides.

'Being under twenty-one at the time I received a letter from Pater. Come and see my roses before the heat spoils them,' Richard said.

Olly and Bill exchanged glances.

'When I arrived, I met the family hanging up on the walls as usual. All painted life size. Then Pater showed me his new carpet. In effect the old boy informed me he'd also heard from the Air Ministry. He said they'd advised he needed to give his consent to my joining the RAF before I would be invited to attend a selection board for a Short Service Commission.

When I went for my interview all they were interested in was, which newspapers I read, what games I played, and did I shoot or ride?'

'What about your father's carpet?'

'Oh, that! He'd had it stretched out in a spare room for inspection. Do you think I should walk on it or hang it on the wall? he asked.'

'And your response?'

'I looked at the closeness of the weave, drift at the uneven margins, and told him to walk on it.'

Olly was impressed. 'Wow! Being clever must mean you're shortlisted as an officer for sure, Richard.'

'It would appear so, old chap. Instructions as to what I should take in the way of clothing included my Saville Row dinner jacket. But now I'm thinking do I really want to join the RAF to become a pilot?'

'Aye, but it's a bit late now to change your mind and go home.'

'I suppose it is old chap.'

In some ways Bill thought Richard, an idle young gentleman and a poor deluded bastard. But he liked him.

Chapter 8

PREPARING FOR ACTION SASKATOON JUNE 1943

There weren't any prostitutes as there were far too many enthusiastic amateurs. The popularity of aircrew, who for the most part were from overseas, meant women's wartime morals became easy-here-gone-tomorrow. This created a minor crisis among Air Ministry officials.

'Legislate away,' Richard said. 'As with leases, writs and statutes, all are written to be read but each by those with their own light of self-interest, which might cause any plans outside of training to be tentative.'

'Aye, but what does that mean in English?'

Richard wrung his hands. 'Should there be the slightest chance of getting laid, no matter how important the plan, all bets are off. Carpe diem.'

'What's carpe diem mean?' asked Bill.

'It's Latin, Bill. It means seize the day.'

'Aye, in Latin, that'd be right. So what does seize the night translate into?' laughed Olly.

'Carpe noctem.'

This led to daunting experiences for Bill. 'There's crumpet everywhere.'

'Aye, there is, Bill. Makes for a pleasant change us blokes outnumbered by women. Back home only place I know summat like this, is Nottingham.'

To most young men, the social side would be a pursuit of sex until exhausted, then one more session as they pretty well reached their finest hour. But Bill had a problem. Inexperienced sexually, he was shy around women. Being brought up to respect ladies, he treated them as he would his mum. They ended up on a pedestal, which didn't help him mentally.

'Aye, you're about as confused as a dog with a choice of two bones to gnaw,' laughed Olly.

Bill's competitive nature had him coming to grips with learning to fly.

He understood the need to apply himself to aerobatics, cross-country and suchlike, but this?

Richard smiled. 'Perfectly understandable, old chap, don't worry about it. It'll take care of itself in due course, won't it Olly?'

Olly laughed in agreement but wasn't himself at all sure of the correct protocol.

'Aye, you might be better attracted to the more elusive butter won't melt in the mouth variety of woman, Bill.'

'All women can be wonderful,' interjected Richard. 'They should be soft and molten in your bed. You shouldn't be so discriminating, Bill. You will limit your potential.'

Bill felt uneasy but this was wartime. Anything goes. He took a deep breath and promised himself whatever came along he would handle it.

'What is your favourite food, Bill,' asked Richard.

'I don't know that I have a particular favourite. I like a variety of foods,' answered Bill.

Richard laughed. 'Think of women as food, but they come in different flavours.'

Canada might have been designated a safer zone, but local men had signed up to go overseas and fight in their droves in support of the Commonwealth. A lot of lonesome girls remained, many who were determined to do their bit to support the war effort. They admired these young patriots who when they became pilots would doggedly put their lives on the line to help protect their homelands.

Whenever cadets waved and smiled at local women they seemed eager to return the compliment.

'Aye, if they resent us being here, they hide their feelings well, by gum.'

'A sure sign of their interest in a uniform and a desire to be sociable with whoever is in it,' laughed Richard. Bill began to realise women were often anxious to be seen in the company of a cadet airman going for his wings. His confidence grew.

The shouting, marching and physical training over, Bill's first introduction to flying was both exhilarating and scary.

When he entered the RAF he'd had a secret fear; vertigo. Standing on a chair to change a light bulb engulfed him with giddiness and panic.

'Don't worry, Son,' Fred had said, 'It's not the fall that hurts you it's the sudden stop at the end. Not likely you'll get killed falling off a chair.'

What would he feel, he wondered, when he left the ground and flew?

It started with parachute training, which unsettled Bill. He went weak at the knees but he learned to pack his own parachute and how to put it on.

The Chief Ground Instructor was a rude man. They agreed he was nothing like Richard's obedient servant. What he said went.

'If you ever have to jump, Cadet, count three slowly after leaving the aircraft, then pull the large ring hard.'

Bill vowed after his first practice jump never ever to use it and never ever to jump out of an aircraft.

Richard took him aside. 'You might want to reconsider your strategy, Bill. One day it might be important to jump.'

'All right. If flames are licking around my arse, I'll jump.'

Richard shook his head sadly at Bill's negativity.

Tiger Moths were used because they were easy to fly and extremely safe.

Issued with basic flying kit, helmet and goggles, overalls and a one-piece padded outfit known as a Sidcot suit, they were almost ready to start.

The Sidcot suit was named after the First World War pilot, Sidney Cotton, who developed it.

Bill was assigned to flying instructor, Flight Officer (FO) Barnaby. A kind man of middle years, he treated the cadets as though they were his own sons. He spoke to Bill from day one, as if he was a young man of limited intelligence.

'First tip, Cadet, is never eating a large meal before take-off. Second is losing those sunglasses. Sensitivity to bright light or not, the relationship will be short-term,' said FOB.

Then he showed Bill over the aircraft, the joystick and what equipment it moved.

'You'll see how when you push forward the nose of the aircraft lowers and you dive. When you pull back the reverse happens, you climb.'

The purpose of the rudder bar and the effect of lift by airflow over the wings made sense to Bill.

'In the event of engine failure it's an aircraft renowned for its gliding characteristics,' said FOB.

On his introductory flights Bill was a non-fare paying passenger and sat like a dummy in the rear seat where FOB could observe him in his mirror. Trussed up like a hen about to be roasted this was the first time Bill had ever flown.

Ground crew swung the propeller and the engine started. It was incredibly noisy.

They began to move at a snail's pace. Rattled and vibrated their way across the bumpy grass then gathered speed. FOB monitored Bill's progress in his mirror.

It was awe-inspiring. Bill had never ridden a fast motorbike, but as they gathered speed he likened this new sensation, to being on one. The wind in his face was unrelenting.

Suddenly the bumping ceased. Bill felt the tail lift before they surged upwards, then in a second they were airborne. Time stood still. He felt as if they were hovering but when he looked down they were flying. The ground continued to fall away and Bill realised he was looking down on roofs and other planes beneath them only seconds before had been beside him on the ground. Bill felt the take-off had been a breathtaking experience.

Bill had no fear of heights. He could focus downwards through thousands of feet of space from the open cockpit without any trepidation, which led to him becoming over excited.

'It's exhilarating, Sir,' shouted Bill down the speaking tube.

'Yes, Bill, it can be. Now you see why you put your goggles on tight. It's important. Everything has to be tight, then tighter than tight,' shouted back FOB.

The experienced pilot put the plane through its paces to demonstrate the cool handling features. It was then Bill realised he had a problem. Airsickness! His experience on the ship should have been a prelude as to what to expect. Bill was as poorly as a sick dog.

FOB was not in the slipstream but next time up involved changes.

'Spitting into your goggles to prevent them from fogging up is important. Preferably in one smooth action prior to starting the engine, rather than as an afterthought with the big front fan at full revs. Oh, and remember not to eat a big meal and drink this mixture twenty minutes before flying.'

Bill was handed a bottle of thick white chalky liquid.

Generally the instructors were helpful and right, thought Bill.

'Only when the manual is as big as the kit do we have a real problem.'

'But what do the asterisks mean?' asked Bill.

FOB looked serious as he took the manual from Bill's hand and thumbed through it. 'There are always asterisks, Bill. Always! Remember that. Put another way, it's the small print that holds the secrets.'

The entire experience was surreal and inspired awe beyond Bill's wildest dreams.

'Unlike in the movies with Betty Grable and Spencer Tracey, this is happening. People out there will be seriously planning to kill you.'

Bill instantly felt alone and horribly scared.

Sensation of speed when flying low exhilarated Bill. Higher up it seemed as though he hardly moved. Sense of speed vanished. If only he could turn that bastard big fan away from his face.

G—force as he dived, knocked the breath from his body. It felt as if his innards were squeezed out through the back of his neck. Bill found it hard to suck in air and learned the value of oxygen.

As the plane turned upside down he experienced a helpless feeling of hanging around in harness. *Oh, fuck! Thank Christ for the harness!*

He tried in vain to tuck his knees up under his chin while his testicles reminded him who the boss in this partnership was. As his disorientation became worse, the harness-thing definitely was a first up task even prior to spitting into the goggles.

Next day they continued with rolls and through to upright again. He learned to cope with the additional handy skill of staying focused through blurred vision and resultant nausea.

'Handy holiday hint number one,' shouted FOB. 'Remember to vomit to the left or right. Opposite side from the airstream.'

Bill only had it wrong the first few times. Aerobatics was constrained to effective flight control and how to evade or attack enemy aircraft, practice interceptions and complete cross-country exercises. None of the cadets knew where they might be assigned. Some would become fighter pilots. Others might fly heavier aircraft like bombers. A few sadly, were destined to the rear of the kitchens.

The job was to learn, grasp and retain knowledge, while leaving others in his wake, but not to become an idiotic thrill seeker. Bill's constant goal was to avoid failing. At all costs not to be reassigned to peeling those wretched potatoes. Bill saw each step forward as a step closer to his coveted nineteen shillings and sixpence a day.

A new way of life began for Bill. He was taught how to climb steeply without stalling and what the hell to do, when he did. His first stall terrified him.

'Throttle back, get the nose up, easy on the stick,' shouted FOB.

Bill noticed the speed drop off rapidly, the noise quietened and the controls grew sloppy. They shuddered and rolled almost upside down. *Oh, fuck!* Then with ease they resumed straight and level flight. *Phew!*

How to determine a safe height from which to dive and withstand the G—forces without blacking out. How to peel away fast to one side and dive. Or to roll in an emergency to make him a difficult target but twirls and victory rolls were out, as was any reckless low flying or showing off.

'Keep your feet on that bloody rudder bar!' shouted FOB.

But nothing was pretty or clean. Planes were streaked along their cowlings, blackened from leaded petrol. Everything was covered with grease, oil and what looked like dried vomit. Scuffmarks from pilot's boots had worn away paint on most upper surfaces of the wings with telltale damage in the more trafficable areas. Bill screwed his nose up. 'The constant smell of leaking aviation fuel is enough to turn a vulture off carrion.'

'Aye, nauseating enough for you but I care naught for their no eating before flying rule, by gum. I get hungry before I fly.'

Bill learned the flying process was not difficult. Soon he was at the controls but the instructor was always in his ear.

'Pull that bloody stick back! Okay. Let's do another circuit. You'll land this time, then take off again—what did I just tell you about that bloody stick?'

Bill was getting the feel of a Tiger Moth.

After seven hours he went solo and FOB was delighted.

'If you hadn't gone solo by ten hours, Bill, you'd have got another instructor in case of a clash of our personalities. By sixteen hours Chief takes a look at you and your progress. After that it's usually goodbye.'

Circuit after circuit Bill flew monitored by FOB from the ground.

FOB was pleased. 'You've done very well, Bill. Keep this up and you'll be my star pupil.'

They had Sundays off and in between learning to fly there were social outings, for the most part dances held in local church halls. Courtesy of various groups intended to make young cadet pilots feel more at home in this strange but friendly land.

In a corner of one such hall sat the latest efforts of the local knitting circle, which included colourful tea cosies. Bill was reminded of Lily and without warning he became down in the dumps and homesick. Being *godly*, liquor was not involved but cadets hid their stashes well.

The official program included lots of hot sweet tea, coffee and soft drinks but with the added attraction of crumpet galore. Richard was in the know and had big news to share with Olly and Bill.

'Word is, chaps, there will be a few nurses from the local hospital at our next dance.'

Bill was confused. He didn't understand why the fuss.

'Surely, you don't think we'll need the attention of a nurse? We're not likely to injure ourselves at a dance, are we?'

Richard roared with laughter. Olly cracked up until his sides ached.

'Nurses have a reputation, Bill. They are so disgustingly clinical and no peccadillo is too extreme,' explained Richard, as he wiped his eyes with his handkerchief.

Bill's expression gave nothing away.

'Aye, they're naught but bikes, Bill. Like to be ridden.'

Even now no answering sparks from Bill.

'Sex, we mean, Bill, for Christ's sake, sex,' offered an exasperated Richard.

'Oh,' said Bill.

'Nurses have to work alongside naked men twenty-four hours a day. They're used to bodily functions, penises, blokes' erections, masturbation, those sorts of things,' explained Richard.

Olly took his illicit hip flask from his pocket and gave it a sideways shake.

'Aye, this might help summat too. Naught but limber them up a bit.'

'Use booze to break the ice you mean. Splendid idea, Olly. At the risk of complicating what has always been a basic activity, some women at these socials might have greater expectations than we do, but we might as well enjoy ourselves,' laughed Richard.

Added to diets of adventure was an eager anticipation of danger that pervaded everyone's daily lives. Newspapers reported thousands of tons of ships being lost in the Atlantic; non-combatants were being targeted in bombing raids in England, and troops everywhere were being slaughtered. Countries were falling like ninepins and the yellow peril was everyone's secret dread. Always there was something important to worry about. Talk by and large was of death and misadventure somewhere, and most people combated this by living for the moment.

'Aye, Bill, if you have money, better to spend and enjoy. There might naught be a tomorrow.'

No one wanted to admit to being scared.

Richard looked serious. 'You realise it's improbable the three of us will survive this awful war. Chances are one, or none of us will make it through. Fate decrees it.'

'Aye, thanks. You know how to dampen the mood and spoil a party.'

That less than half might survive was a sober thought. As with Canadian men who fought overseas, few would return home.

'Fate is no more a bound servant of the deserving man than it is of the undeserving,' continued Richard. 'Besides destiny of people is for the most part decided by a few men in a room. That is how the world is changed, by the stroke of a pen.'

None the less the menaces of war meant risks were being taken by non-combatant civilians everywhere and that included pretty women who would as a rule exercise caution around men. They often threw care to the wind as few put off until tomorrow, what they could do today. To have their way with women, men promised the earth.

The word 'love' was bandied about a lot. Pledges were made and promise to write letters ad nauseam. Guarantees to return for loved ones as soon as hostilities were over helped some to lower their expectations and several women were none too discerning about allowing themselves to be used sexually.

'Aye, some will be in and out like a fiddler's elbow,' laughed Olly.

Chapter 9

BACK TO SCHOOL

Training courses were theory and practical mixed. Simulation included working in pitch- black conditions. Olly couldn't see his hand in front of his face. 'Aye, it's the sort of inkiness smothers life itself.'

Richard frowned. 'Darkness might be our best friend if we're trying to evade Jerry.'

Bill was thoughtful and surprised. Once his eyes became accustomed he could see quite well.

They'd listen to a radio expert, then study celestial navigation or try to understand how flaps work to reduce speed. Various problems came up including emergency situations. Some classes were intended to monitor what each individual does in time of varying degrees of stress and crisis.

'Naught need to monitor me,' laughed Olly. 'I get scared shitless in any crisis, by gum.'

Captain Hawkins, an expert on bombsights, communications and radar was introduced to the class. 'Today I will talk about radios for wireless operators.'

He spoke knowledgeably about advanced communications systems but apologised to the group as he felt the need to generalise. 'None of you as pilots elect are aware of what machines you might fly after qualifying. Knowing what squadrons you will be assigned to would make life simpler but I suspect no one has that information.'

Captain Hawkins shrugged, as if with some discomfort. 'Even if they did, they would be unlikely to tell us so I need to give you a comprehensive overview. You shall have to study everything I'm afraid, in the hope you remember later what you need to know.'

Another cadet, Duncan, looked uneasy, almost shifty.

'He looks like a parolee,' scoffed Richard.

Duncan too was born into a life of privilege with multiple chimneys, but he lacked the common touch of Richard. Another reason Bill had grown to dislike him was he kept taking the piss out of Bill's surname, Honey. Bill tired of being referred to as 'honey sweeter than sugar,' and 'you're not as sweet as your name, plus 'come here sweet, Honey thing.'

He had copped ridicule at school but hoped, in the company of mature men, it would cease.

Duncan had a question. 'Please, Sir, will we get much practical instruction? I never seem to learn as much from a book.'

'You should try opening one,' sneered Bill, who was in no mood to be charitable.

The class laughed. Captain Hawkins thought the comment appropriate and smiled.

'As much as possible, Cadet, yes. I'll start with our Canadian aircraft and those built for service in the Far East. They have radios made in America. We can show you one of those tomorrow. They're special in they provide radio direction-finding, as well as having voice and Morse capabilities.'

'What about the Oboe system, Sir?' asked Richard.

'Very good, Cadet, you've been doing your homework. Truly, I am impressed.'

Richard beamed with pride.

'The Oboe is an accurate navigation system but has limitations. It can only handle one aircraft at a time and is fitted to faster and more manoeuvrable, aircraft. Pathfinders and Mosquitoes. It's unlikely any of you will fly the Avro Manchester twin-engine bomber. It's been around for about two years but has had nothing but problems.'

'Why, Sir?' asked Duncan.

'Powered by two Rolls Royce Vulture engines, it has proven to be completely underpowered and extremely unreliable. Not to mention its airframe defects.'

'Word is anyone who flies them is a sitting duck, Sir,' added Richard.

Captain Hawkins stared at the floor. 'Quite so, Cadet, they share a similar reputation to the outclassed Blenheim.'

There was a palpable unease in the room. Many shuffled, some groaned.

'But not to worry, chaps. On the credit side, it paved the way for the Avro Lancaster, which in effect is a redesigned Manchester. We now have brand new four engine bombers to include the Stirling and the Halifax. The new Avro Lancaster has longer wingspan, takes two extra engines with far greater bomb capacity. If any of you end up flying Lancs, as I suspect some of you shall, you won't have the Oboe.'

'Why is it called the Oboe, Sir,' asked Duncan.

'Anyone, Class,' responded Captain Hawkins.

'Because it sounds like the musical instrument, the Oboe,' offered Richard.

Captain Hawkins beamed. 'Quite so,' he moved casually about the room as he talked but his eyes darted everywhere. He missed nothing.

'The GEE is a radio receiver by which a navigator is able to plot his course on a grid. GEE stands simply for grid. It functions by synchronised pulses transmitted from England.'

He continued, intermittently turning to face his class, between scrawling notes on an enormous blackboard.

'Please, Sir, should we be writing any of this down?' asked Duncan.

Captain Hawkins turned and stared rigidly at Duncan. His classmates giggled nervously.

'Aye, Duncan, might be summat worth considering,' Olly said quietly.

'The GEE has a range of 400 to 450 miles, which is limited by curvature of the earth. You calculate your position based on time delay between pulses but it is not a pinpoint system.'

For two hours the cadets painstakingly familiarised themselves with intricacies of the GEE and the manner in which pulses functioned. Duncan tried to concentrate but found the effort beyond his attention span. Bill became engrossed and asked salient questions.

Captain Hawkins looked at his watch. 'Okay, enough of the GEE. I want to tell you about the GEE H. It's similar to the Oboe.'

Duncan looked bewildered.

'Don't become confused. It's easily done. With transponders on the ground the GEE H allows more aircraft to use the system simultaneously. Any aircraft fitted with this are usually marked with two horizontal yellow stripes on their tail fins. H2S is a ground-looking navigation come radar system but we fear it might be homed in on by German night fighters. It has to be used with discretion.'

'What does that mean, Sir?' asked Duncan.

'Don't turn it on, old boy,' offered Richard.

'Yes, quite right, Cadet, might be expedient.'

Some cadets laughed.

'But we will have something new soon, the Airborne Cigar. Has anyone heard of it?'

Bill was not alone when he shook his head.

'Nick-named the ABC it will have three aerials.'

Another bloody calamity, thought Bill.

'In a Lancaster two will stick out at the top of the fuselage and one will be under the bomb-aimer's position. The ABC will be unique as it will be used to jam radio to German night fighters but it's in the prototype stage, I'm afraid.'

Captain Hawkins began to gather up his papers. 'You will be tested on general information but next week will be more hectic when we cover the H2S, the Fishpond, and the Boozer.'

Everyone started to laugh nervously, a fact appreciated by Captain Hawkins who was grateful for some light relief. 'Boozer is a radar detector but it might be more disconcerting than useful. Its system of lights mounted on the instrumentation panel are supposed to glow bright when tracked by Wurzburg ground radar or Lichtenstein airborne radar, but we fear they are often triggered by false alerts.'

'Another one to keep turned off, Sir?' suggested Richard.

'Yes, quite so, Cadet, those among you who have been Eagle Scouts might have minds like a compass but for the rest of us, it's hard work I'm afraid.'

Bill felt confident, as he'd been a Boy Scout. He felt he no longer needed to ponder his fate with these exams.

He twiddled with his pencil as he studied his notes. He could do this, he was sure he could. More than ever he had no intention of being scrubbed as a pilot.

Richard quoted Virgil, *yield not to calamity but face her boldly.*

Chapter 10

AN INTRODUCTION TO SEX

In Bill's defence not everyone fared well in the classroom, or the theatre of love. Then as today, some might have had more fun paying speeding fines than attending one of those church hall socials.

Richard chided Bill. 'It rather depends on your attitude. Maybe you crave a good foot rub, old boy,'

'Aye, or he's got his eye out for someone with summat special,' smirked Olly.

'Someone who might re-fill his pipe, and tamp it down for him, the way he likes it,' offered Richard sarcastically.

Olly and Richard were taking the piss.

At times like this, Bill felt like a salmon attempting to swim upstream but he realised Olly and Richard's intentions were friendly enough.

No such qualms from Olly, Richard and some of the others. Most wanted to get in for their chop, whenever and wherever they could. Marooned by circumstances beyond their control, they were determined to make the best of it and dry hump anyone in Canada who would let them.

Social dances started out being supervised. Once underway the older folk departed with a promise to return later to help close up and clear away. But they rarely did. Richard behaved in an exemplary manner but gave Bill a broad wink.

Any girls in a serious relationship, those who didn't want to mix in, or anyone feeling 'poorly' because of their wrong time of the month kept 'watch,' helped make tea or wound up the gramophone and changed over the 'seventy-eight' records. A favourite was Glenn Miller.

Married women seldom attended.

Some cadets were content to sip tea and chat with a girl, any girl. Others expected more.

Initial introductions and preliminary dating rituals completed, girls selected their preferred partners for the evening. Music slowed, lights dimmed and tentative face sucking precluded more serious necking, untidy dishevelment, petting and states of undress even as far as point of nakedness.

'Whatever a woman's self-consciousness feels fit to overcome,' beamed Richard.

On first dates it was for the most part chitchat followed by carousing.

After their first uneventful church hall dance Olly and Richard exchanged notes and tried to guide Bill.

'There's a lovely blonde nurse named Anne. Olly and I both have our eyes on her.'

'Aye, she's a goer for sure. I like her. She's a blonde with long hair, big tits and a big arse. Built for comfort not speed, by gum.'

'You'd like anyone willing with a pulse,' laughed Richard.

'Aye, but it's hard naught to look at Anne's tits. Especially when they enter the room a fair way before anything else.'

'Out and proud,' beamed Richard. 'It's like her breasts are taking her for a walk.'

At their second church hall dance Anne latched onto Richard as if he was an ex-boy friend back from the war and flirted outrageously with Olly.

She wore a knee length tight chequered skirt and a pale blue silk blouse that couldn't contain its contents. Bill saw flashes of bra and skin between the pearl buttons. Her chiffon scarf tied loosely at the neck did its best to retain some propriety.

For some it was more a case of finding a space that allowed such freedom as only betrothed women might allow. Away from the main floor was in great demand. No one wanted to be caught out if someone in authority decided to come back to check on progress of proceedings.

Couples vied to use dark corners or any area of privacy where they could continue in part or all the way. Popular was anywhere with a lock or a key; toilets, kitchen, or any store room, no matter how cramped or small.

'Aye, anywhere large enough to break wind will do,' laughed Olly.

'Bill, you're yet to realise, women need a reason to have sex, men only need a place,' confided Richard.

'Not every woman needs to have a reason to have sex with a man,' said Anne. 'Sex in itself can sometimes be enough. I know it is with me.'

To Olly and Richard, Anne was now the most attractive woman in the room. About as popular as a naked porn queen handing out free drinks in the Mess.

Punctuated between contented puffs at his pipe this opportunity for Bill to sample delights of an enthusiastic amateur, perhaps even get laid for the first time many miles from home, began to excite Bill. He tamped down the tobacco in his pipe with his thumb, and lit with repeated match strikes before sucking deeply to his satisfaction. As he exhaled Anne put her face closer to the smoke.

'Ahh, I like it, Bill. But it's not as good as this mother's milk.' Anne drained her cup in one. 'Drinking Scotch is like slipping naked into a warm bath.'

The prospect of Anne naked in or out of a bath, any bath, excited Bill. He hadn't had many opportunities of intimacy with members of the fairer sex, but now, he realised, might be his big chance. He took a good look at Anne as she wrapped her arm in Richard's. She was a biggish girl, a bit heavy set, with enormous breasts. But she seemed keen on Richard and liked Olly. Bill sensed it was time to go and find his own girl.

'When you latch onto someone, Bill. If you're not sure what to do, keep kissing her. Women like to be kissed. If you kiss her right, one thing leads to another,' advised Richard.

Anne nodded in agreement. 'No argument with that.' She laughed. 'That's usually how it starts.'

Anne was about to let her Veronica Lake hair down. She was almost pissed.

They laughed.

All pretences of etiquette were about to disappear about as fast as a snowflake settling on a hot coal. Bill excused himself and went off to explore the room.

Soon he met a beautiful Canadian girl named Sophie. She worked as a part-time nurse. He shuffled his feet in time to the music a couple of times as they pretended to dance, but as he'd never danced before his was a half-hearted effort. Rather than pay attention to the direction his feet were travelling Bill was more interested in staring down into the acres of Sophie's wide Canadian smile.

She was a slip of a thing, not large or buxom like Anne. Small breasted, but as she moved about they pushed against her blouse and looked inviting. Bill searched deep into her big doe like brown eyes and wondered at her exotic beauty. She had even white teeth, which flashed against her coppery-olive complexion. Some Italian heritage and with and an obvious liking for younger men in uniform by the way she'd taken to him.

Sophie was attractive and nimble on her feet. Occasionally when Bill glanced across at Olly, he noticed he and Richard gave him a thumb up sign behind her back.

Then Richard raised his cup and did an exaggerated pelvic thrust and they all laughed, including Anne. It was as if she couldn't get enough of Richard, but wanted Olly's attentions as well.

Sophie was a high-waisted girl with jet-black hair pinned up in a bun. Co-incidentally the same as Bill's mum wore. He felt at ease and as they shuffled about and laughed together. He imagined this might be pleasant if it moved up a level.

Then he wanted to kiss her. He leaned in to peck her on her cheek, and as he did so, she anticipated his move. She moved her face sideways to catch his kiss square on her mouth. During the kiss she kept her eyes open wide and laughed at Bill's surprise. Ice broken they kissed again, slower than before. She tasted like candy floss. Bill licked his lips and drew Sophie closer to him.

Her scooped dress collar rose up in their embrace and gaped open a little. As she leant towards Bill it opened wider. He imagined being able to rest his hand on her naked shoulder and follow with his thumb the hollow beneath her throat. From there he tracked the line of her breast as a child's eye follows a line of print in a book.

Between kisses he had an uninterrupted view of her exposed cleavage. Her small firm breasts looked larger than before. Another shuffle disguised as a dance and Sophie became impatient. She took Bill's hand and with much sniggering and laughing between her and some other girls, led Bill away down stairs into the basement.

To Bill's surprise others were already down there. Anne's long blonde hair made her look a bit like a film star, but she unquestionably had that please-mess-with-me attitude. As she sashayed about drink in hand, she kept looking for refills.

Wow! Those two are fast workers when it comes to women, thought Bill.

Sophie reached up and took hold of Bill's face in both her hands. She opened her mouth wide and took Bill's tongue and rolled it about with hers. Rather than lose it entirely in the exchange, Bill reciprocated. He had no trouble being caught up in the moment. Each time his hand brushed against her breast, Sophie pulled herself closer to him.

Sophie reached up, tugged for a moment at the comb that held her hair in place. She shook her head vigorously from side to side and laughed as her hair cascaded over her shoulders. She then led Bill into an empty toilet cubicle and snapped closed the lock behind them.

Thank Christ. At least Canadians have lockable doors on their shit-houses.

CHAPTER 11

SEX WITH AN ENGLISHMAN

Inexperienced, Bill became faced with the inevitable. Sophie braced herself, her upper back and shoulders up tight against the lockable toilet door. Her hair fell about her in a cascade of want as she balanced on her high heels and slid downwards. Deftly she undid Bill's belt buckle and then removed her own under garments. Bill realised this basic training might be in Braille but he should find his way.

Sophie prepared to make way for Bill, as she would an experienced lover. But Bill held back. He was completely carried away with her charms and willing enough, but a little unsure of what to do.

Advice from Richard came to mind.

When you latch onto someone, Bill. If you're not sure what to do, keep kissing her. Women like to be kissed. If you kiss her right, one thing leads to another. Augmented by Anne, 'no argument with that. That's how it starts,' she'd said.

Bill continued to kiss Sophie. It wasn't a chore, as he wanted more of her. After a while he began to feel not only her reactions, but his own.

Sophie felt him too. She began to purr, but was surprised he held back. Oh, well, she thought, perhaps he likes to take a girl slow, might be a pleasant change for once. I can go with that.

She whispered, 'Cherie,' and other words in French, none of which he fully understood. Balanced as she was, she shuddered as he slipped his hand gently up her bare thigh.

'Have you come prepared, Bill?' Her words were short and urgent.

'Prepared for what?'

Sophie paused. 'Condoms. Have you a condom, Bill? Now's the time to put it on.'

'Right.' He reached to search through his pockets. He knew he had a packet somewhere. Richard had bought them prior to the dance and shared them around. When he'd handed Bill the packet of three, he said, 'Bill, here is your master key to a guaranteed shag, old boy. Guard it well.' Bill found the packet and started to fumble.

'Give it here,' said Sophie. Passion in her voice replaced with urgency as if scolding a child. Sophie tore into the packet with the nimble fingers of someone more practised at operating a Morse code machine in darkness, than being a part-time nurse.

'Here, move! I'll put it on for you. We haven't all day.'

Lacking was a comfortable double sized bed with cool starched sheets, thought Bill.

Now sheathed up, Sophie had lost her moment. She placed his hand on her one naked breast, then the other alternately. As she did her nipples became erect again. They stuck out like little gun turrets on a fighter plane.

'I've never had sex with an Englishman before,' said Sophie in a hoarse tone of voice.

'Neither have I. Makes two of us, Sophie.'

She laughed a contented deep-throated laugh, which to Bill didn't sound at all ladylike. It was a dirty laugh. Then she arched her pelvis as if to accommodate his entry at a more comfortable angle as Bill continued to fumble and lay against her. Her breath came in shorter gasps.

I'm about to have my first shag, thought Bill.

As he experienced his first knee trembler, his legs found his knees utterly out of control. He gyrated frantically from side to side like a mad thing as he lunged at her. Bill's first poke, or at least his first serious attempt at having one ended prematurely, and not well.

Nothing penetrates fog quite like panic. It set in when Bill's fingers strayed to above Sophie's womanhood and onto where her stomach should have been. Instead of yielding flesh he felt something hard and uncompromising. It threw him completely. There was a substantial hard casing where only pliable flesh should be.

Bill pulled back. 'Bloody hell!'

'If you don't royally stuff things up, Bill, I will see you again,' promised Sophie urgently. But it was too late.

'What's this?' Bill tapped his fingers against the immovable and steadfast object.

There was a pause.

For Bill dire visions piled one on top of another.

'Oh, don't worry about that. I've been in hospital for an operation. It's nothing but a plaster cast.'

This was a revelation to Bill. Discovery of a steadfast plaster cast into which Sophie's entire abdomen was encased, caught him off guard. That this delightful slip of a girl had undergone extensive surgery was one thing but her entire abdomen encased in post-operative surgical plaster was entirely another.

Bill's concern in a flash killed any semblance of romanticism or sexual desire in him. Sophie's enthusiasm never slowed but Bill found himself in a state of undress, deeply embarrassed. Here he was with a beautiful woman who was making a considerable effort to convince him how they should continue and how it would be quite all right if they did. But Bill was fearful he might do her harm.

'I'm sorry, Sophie.' To him, the prospect of her guts falling out of her plaster cast in the confines of a toilet cubicle placed a whole new perspective on his capacity to draw a crowd. He visualised this lovely girl, her innards popped out onto the floor directly at his feet. Sophie sensed all was not well.

'Was this your first time, Bill?' she asked gently.

Bill felt like a country lad, slow and thick as syrup. 'Yes.'

He felt dejected. If he'd pumped his heart full of high-octane fuel it couldn't have beat any faster than before he'd discovered her plaster cast. The look on Sophie's face did nothing to cheer him up. He'd heard stories throughout his growing years of conquests and was down in the dumps. His big opportunity with beautiful Sophie had not ended well.

Before they left the cubicle, Bill complied with her solemn wish to commemorate for her, what had been a more memorable if not exalting experience than for him. He wrote his name and date in a small-nominated space on her overcrowded but popular plaster cast as it was soon to be removed.

'If we don't succeed, we run the risk of failure'...Dan Quayle

Chapter 12

YOUNG COMRADES

Two weeks later Olly was in hospital.

Bugger!

He'd been bitten all over his backside by mosquitoes. Caused in part by laying face down exposed, having sex in a wheat field, which was a popular venue with cadets close to Base.

She was an attractive half-caste Inuit girl called Jill he'd met briefly at a church hall dance. Jill had indicated to Olly in no uncertain terms her willingness to let him, and him alone, make love to her. It had seemed too good to be true.

'In hindsight, perhaps they should have found a room,' Richard said sarcastically.

Bill remembered Jill. She'd been distinguishable from a distance. Her black bra showed through from under her flimsy white blouse. The cut clung to her ribs, fell snug across her flat stomach. Her stark white teeth, sultry dark eyes and head of flowing raven black hair. Set against her thin strong face and iodine coloured skin she'd been hard not to notice. She'd sat by a slatted window quietly reading a magazine. Defused light reflected off her dark features. She was pretty but had spent too much time in the sun. Her darker colouring accentuated her looks but made her look too pruned, almost a reddish purple.

She looked out of place at dance socials because of her constant thumbing through magazines.

What cadets hadn't realised was, she wasn't reading at all, only looking at pictures. Her English was as poor as her inability to read. Only thing Bill hadn't liked about her was she smoked every cigarette offered to her.

Bill looked around with a slow lizard blink. Although he tried not to show any jealous feelings towards Olly, deep down he wished he'd been with her in the wheat field. There was something about her innocence and calm that put her on a pedestal.

He stood beside Olly's hospital bed and nervously fingered his pipe in his pocket. If only they would let him light up. He felt twitchy in hospitals at the best of times. The scent of disinfectant seemed to permeate everywhere about him.

Olly lay face down. He looked drained and blanched as if someone had knitted him and drawn the stitches too tight. His normally tousled red hair, thin on his scalp, was laced with sweat. In the last few hours since Bill had seen him, Olly had visibly aged. He bore no resemblance to the keen young man who a few hours ago had been knocking back alcohol and shagging himself silly.

Strapped to the cheeks of his bare arse were two massive reflective boxes housing hot light bulbs. Extreme heat generated by the device was intended to draw out any infection from the severe bites. Bill leaned across the bed to get a better view and marvelled at the incredibly consistent group pattern of the mosquito bites.

'Bet it hurts. Looks like both cheeks of your arse have been a rifle range target, Olly.'

Olly pretended not to hear his friend. Instead he turned his head away and gently sobbed into his pillow. The throbbing sensation of his discomfort bordered on unbearable.

A buxom nurse with short dark hair in an impeccable starched white uniform appeared carrying a tray of instruments. She turned to talk to a doctor who had stuck his head around the door then eyed Bill from the other side of the bed. There was something extraordinary about her and Bill felt he should know her from somewhere. He was unable to take his eyes from her.

She looked familiar and kept smiling towards Bill. He felt embarrassed. He couldn't quite place her.

'I bet you're pleased your friend's in here and not you, Bill?' said the nurse.

Bill was taken back. How did she know his name?

She noticed his surprise. 'We know each other, Bill. I'm Anne.'

Bill stared at her. 'But you can't be, Anne. Anne's a blonde. She has long blonde hair and she looks different!'

'I'm impressed, Bill. Yes, I was blonde. Past tense. During the week I work as a nurse and I'm a brunette. At weekends I'm a blonde. I wear a long flowing wig.' And then she lowered her voice as if to divulge some deep dark secret, 'Bill, that's because they say a blonde has more fun.'

Bill didn't need to be conversant with Albert Einstein's complex mathematical formulas to work this out. Last time he'd seen Anne she was a platinum blonde party girl. Now she was a short dark haired professional. Of course she looked different without her wig and in uniform. But not even drab neck-high garb could entirely hide the best tits in Saskatoon, thought Bill.

'You're right, Anne. I am pleased. Relieved I'm not in Olly's shoes now although there were times when I wouldn't have minded trading places with him.'

Their conversation halted when a middle-aged man entered in a white lab coat with a stethoscope around his neck. Bill stood back from the bed and offered to leave.

'No, Cadet,' the doctor said gruffly. He eyed Bill up and down. 'Better you stay and watch the treatment and tell others. It might discourage a few more of you young idiots from acting recklessly, as this cadet has done.'

The doctor glared at Bill. 'Your friend is suffering not only from the pain of the lamps but the debilitating and depressant effect of the drugs we've pumped into him. He's about to learn his sojourn in the wheat field bestowed upon him a more personal memento of their union. Aren't you, Olly?'

Olly groaned and tried to force his head deeper within the pillow.

'He's got a severe dose of the 'clap,' said the doctor, flatly. 'Not a happy ending for you, Olly.'

Bill was shocked. Now he had an entirely new perspective on relationships with pretty but permissive girls in wheat fields. Jill had appeared so demure, so innocent, a lady who could have been a one-man type woman.

'Venereal disease is viewed somewhat dispassionately by the authorities,' continued the doctor in a serious tone of voice. 'Not an acceptable occupational hazard, Olly, is it? Well, not according to the tried and tested manual, it isn't. Sexually transmitted diseases are frowned upon but now might be a bit late to lecture you, Olly. You should know to wear a condom.'

* 4th Footnote

The doctor held an instrument up to the light. It looked a bit like a closed metal cocktail umbrella. 'I'm about to pass this instrument into the shaft of Olly's penis,' explained the doctor. 'Withdrawn slowly, it's designed to open and scrape any deposits of infection or pus from inside of his shaft.'

Too bad for Olly who was awake and sober.

'Aye, Doc, but surely this is naught but a deterrent than summat of a treatment,' whined Olly.

Bill felt faint. He sat down heavily on the corner of the bed and clutched his private parts as the room swam before him. He felt he needed oxygen. As a psychological measure it might go a long way to ensure Olly and others stay out of the wheat fields at night but,

'*Oh, fuck*!' Bill shouted in his head.

Olly began to sob. 'Aye, but it's a pity you can't catch up with culprit who gave it to her in first place,' sobbed Olly. 'Why me, Doc?'

Enough to discourage many testosterone filled cadets not to be reckless, the mere thought of it had already turned Bill a brighter shade of pale. He felt beads of cold sweat between his shoulder blades and his palms were clammy.

Olly promised the doctor and any god who cared to listen how he would wear a condom in future, but they were unimpressed. Attributable to the amazing, shrinking evidence of the existence of God, perhaps, thought Bill.

Irrespective of the doctor's wishes, he left the room.

Two days later and things were almost back to normal for Olly. In spite of everything he wore a pained expression on his face and it hurt like hell to pee.

Whenever Bill looked at Olly he felt panic leach up his spine from the image in his mind of that demonic metal umbrella.

In the hope Olly might benefit from fresh air, Bill suggested they go for a walk and while at it, perhaps shoot at something with their easy to clean guns.

Richard was unimpressed with the concept and stayed back to read and write letters home.

They skirted around the wheat field looking for rabbits to shoot at but found no sport.

Olly pondered and waxed poetic. 'Aye, wheat still twitches and rolls whenever wind blows but summat about this place naught looks the same.'

'It's a bit like attending a re-union, when no-one else shows up.'

Bill's humour was not entirely lost on Olly.

'Particularly with your mozzie protectors on,' pressed Bill. 'By the way, how is Jill?'

'I'm naught sure. She's getting treatment at hospital arranged by Anne, but she looked really upset. Her church thinks it might be best if she moves away.'

'That's a shame. Have you discussed it with her?' Another bloody example of the church's cold charity, thought Bill.

'Aye, I've tried, but it's naught easy. 'My French is non-existent and her English is summat poor to say the least.'

'That does tend to make conversation difficult. But I suppose in the wheat field she didn't need to say much.'

'Aye, you're right, Bill,' Olly grinned. 'Least of summat on our minds, was listening to what I had to say.'

'Everything looks different from a dissimilar perspective. Like with you when you're standing-up, as opposed to lying down,' Bill said dryly. 'What about Anne?'

'She's as prickly as a holly bush, giving me a bit of a wide berth. Can't say I blame her. Me getting the clap like that. But I think she's seeing a bit of Richard.'

'Which bit?'

Chapter 13

LOST

They weren't pilots yet and even now could fail. If it had been easy for Bill to get in, he'd best bear in mind how it might also be easy to leave. He took deep breaths and suppressed his negative thoughts. Aside from the war, which seemed a long way away, for him the day positively chimed. He felt sure it would all come together and the loose ends join up at the middle. Pleased and confident with himself, he sported a new pipe with a cherry wood bowl from the shop at Base. But he was to learn there was a fine line between using his initiative and blaming everybody else for the lack of it.

'It's an intriguing business this learning to fly, Olly.'

'Aye, it's naught but a fine line all right, summat hard between being wrong and a visionary,' Olly frowned.

'But perhaps you need to be an idealist to see it,' offered Richard. 'I see General Dwight Eisenhower is named commander of Allied forces in Europe.'

Olly and Bill were only vaguely aware of who Richard was talking about.

Unofficial bombing run practices were encouraged for a pilot to better understand how, if they dropped something from a great height, it ended-up way, way off the mark. But not much time was spent on dropping things.

'Pilots assigned to fighter planes don't need extensive throwing practice and bomb-aimers are trained elsewhere,' said FOB.

Instant dismissal followed any unnecessary low flying or attempts to show off by flying under bridges or the ultimate statement of ability—through open hangars. To land, other than in an extreme emergency, in a wheat field or on a road was not tolerated by His Majesty.

'Aye, better keep out of them wheat fields, lads,' offered Olly.

'Clear as mud. Where's that hangar, the one closest to a church?' laughed Bill.

Their mind-numbing training plan included steadfast rules such as never to swerve on the runway in an attempt to avoid hazards, no matter what their size.

'Be they as small as a fluffy tailed rabbit or eye contact with a horse, you go straight on through,' instructed FOB.

'Once you're given clearance to take off and 'it's a go'—you go!' reinforced FOB.

'If anything does get in your way once you are committed speed-wise, you do have to pass through it and hope like hell it won't hurt you, or do too much damage to your big fan up front.'

'And while pledging allegiance to your preferred creator,' laughed Richard later in the Mess.

Bill's major and unfixable concern was how to turn his aeroplane around on the ground. Quite effortless in the air, but on the ground, physics and manuals kindly printed in the King's English, didn't help him. Not even the parts with asterisks. FOB shook his head sadly.

'You can't always travel in a straight line, Bill.'

'But I want to be a fighter pilot more than anything, Sir.'

'You might be if you learn to handle the bloody thing on the ground.'

Later in their Mess, Richard offered a suggestion.

'This task could need some good old fashioned customisation.'

'Aye, if generations of hardened piss-heads can do it, why not him,' laughed Olly.

Bill felt like he was wallowing in his own bodyweight of shit. No matter how vigorously he turned the rudder, kicked the ailerons and over revved the engine past point of embarrassment—nothing.

He was convinced he was doing everything he was supposed to do. With their grass strip progressively shorter at each attempt Bill was beginning to think he could best support the war effort, by not being there at all. A small injury and admittance to a nice hospital with pretty nurses sounded preferable.

'What level of hell is this, Olly?'

'Aye, summat's not quite correct, it's as if your plane naught wants to play.'

Then a bolt of brilliance made Bill feel confident. Well away from prying eyes, he undid his Sutton harness and with engine ticking over and brakes locked, clambered out of his cockpit and picked-up the tail. With the mustered strength he only ever thought possible by his eight-foot Hun nightmare fiends, he hastened little fairy steps to skip the bloody thing about-face a full one hundred and eighty degrees.

Ah, customisation at last.

Unfortunately there was a witness, Duncan. On return to their Mess Bill copped tremendous criticism from the entire group.

'Aye, harsh but fair, by gum.' Olly saw his opportunity to repay his friend for his wheat field comments.

Un-abating ridicule of Bill's plane turning debacle from the team was severe.

FOB scratched his head. 'Most pilots have a problem mastering the art of something or other. With some it's landing as they bounce like kangaroos hopping about the place, with others it's taking off as in gravity thou art a heartless bitch!'

'Aye, it has manifested itself for you, Bill by your inability to turn the wee plane around on the ground,' smirked Olly.

At night Bill lay awake in his billet and worried. He began to prepare his daily plan to peel potatoes in the Navy, on the rolling deep. Would he survive this weeding out process? Might his plane-turning problem be his undoing?

He would have gladly taken up any colonial appointment offered him. Evidently the instructor's advice, useful to everyone else, had struck no answering spark of revelation within him and yet on occasions he'd fluked it. Otherwise he would have been unable to go solo. If only he could master this one thing.

Feverishly Bill hung on to his false smile, but he felt he'd blasted his career as a pilot even before it had started. It was Richard's air of ultimate confidence and encouragement that gave him the determination to push aside his handicap of negativity and try something different.

The agony of the group humiliation passed, followed by some unofficial and private tuition instigated by Richard, but quickly taken up by Olly and others in support of Bill.

'The trick, Group,' explained Richard as he warmed to his task, 'is for each of us to bear one another's burdens. Whenever one of us stumbles, it's like in *Alice in Wonderland* when she falls down a rabbit hole. Often it's a bit of strange luck makes things roll. Bill, you need to feel like a pilot, and not a hobbyist. Follow the rabbit.'

'What shit's he on?' someone asked.

The group nodded although no one had a clue what Richard meant but Bill was rapt. Bill understood. He could see an opportunity.

'Anyone lucky enough to be taught by Richard after the war is over will be most fortunate, Olly.'

'Aye, summat like a mentorship can be important, Bill. I knew a chap at Veterinary College. He was a godsend, by gum.'

Bill was convinced Richard was his lucky break at a time when he needed it.

'Having someone in your corner, to champion you, makes a big difference,' Olly.

Snooker tables end to end became an impromptu runway and a wheel chair from first aid an improvised Tiger Moth cockpit seat. Both handily situated next to their bar. No wild yearning for the unattainable for Bill. What he wanted was to turn his plane around on the ground, but although everything looked about to fall into place, he, upon falling down the rabbit hole, like Alice, had to decide to follow the rabbit. Not everything can be left to other people he told himself.

'Making mistakes is part of being human. Without them we will not grow,' Richard added.

'Aye, nor have anything to laugh at,' smiled Olly.

'Man is born to trouble, as sparks fly upward,' continued Richard unfazed. '

Bill's colleagues were convinced he had more on his mind than in it, which presented challenges to their patience almost beyond endurance. Richard took charge.

'Take a deep, deep breath. Slow down and we'll go through it all again, one more time.'

Everyone groaned.

'Then one more time again, and again, then slowly again from the beginning. Just because the problem's turned out to be bigger than King Kong's scrotum, is no reason for us to give up on Bill. Nil bastardi carberandum!'

'What's that mean?' asked someone.

'Don't let the bastards grind you down,' answered Richard coolly.

'Sweet bloody Bill Honey's a dick-head,' said Duncan.

'That, or he's an ape with limited intelligence and a sweet tooth,' chided another.

'Talking of apes, it's a bit like asking a monkey to paint a Renoir,' said Duncan with a smirk.

Richard snarled. 'It's also about trying to find a practical trajectory between your ears, Duncan.'

'Aye, proof Ballsy-monkey Bill is back on track by gum,' Olly said supportively.

'Painting-up a storm, if not a Rembrandt,' laughed another.

'Give him a jar of honey,' called out Duncan. 'Then, at least he might catch up with the rest of the team.'

Bill grinned his most confident smile, impatient to put his newfound knowledge to the test. Success sparkled on his horizon. He wanted to seize it, take hold.

'Carpe diem,' Bill said quietly.

Rejection and failure no longer loomed. A few more whiskys, an animated fall or two from their bar stools, and the stage was set.

'Perhaps you have a drinking problem, Bill,' Duncan smirked, 'don't you think?'

'You're a Mormon, Duncan. Next to you, everybody's got a bloody drinking problem.'

Out they stormed into the night in search of any Tiger Moth fuelled up, ready to go.

All except Richard who wished Bill well in his usual dignified manner.

'I'm convinced you will have no further problems with your conquered task, Bill.' Richard retired to write letters home and adore himself in his bunk.

Rules on Base were strict about being off your face.

Olly was concerned for his friend Bill as borrowing unassigned aircraft after hours and flying them without due care and attention was right up there with dismissible offences.

Bill pointed out when called to account, with the helpful intervention of Richard, how he hadn't actually flown a Tiger Moth without permission.

'I only went tats with it, Sir. I taxied it. Turned around and around on the ground,' said Bill proudly.

The powers were well informed by Cadet Duncan who became ostracised by their group on account of his dobbing in nature.

'Aye, you're naught but a copper's nark, Duncan,' accused Olly.

After that the powers changed the wording in their rulebook to include; 'no sitting in the cockpit of a Tiger Moth with the engine on without written permission.' Bill knew he was lucky not to have been thrown out on his ear.

The powers appeared unmoved at his extra curricula efforts to fully master a turn. But among instructors, like FOB, Bill had allies he never realised, men who batted hard for him to succeed.

'Try to pace yourself a little more, Bill,' advised Richard. 'You are a talented and gifted young man deserving of a wider audience, but you don't always show others your best side.'

As a result of Richard's guidance, Bill had mastered the correct technique of being seen to conform. Able to remain as intended inside his aircraft, even through the most complicated turns on the ground.

Also, his spoken English improved, made easier by slower speech.

Bill had not only achieved the ability to turn the aircraft on the ground, but had formed extended friendships with Richard and Olly beyond drinking and whoring buddies and had the potential to outlast time. Little by little he felt more at ease.

'Try to pace yourself more, Bill,' Richard advised. 'I've told you before you don't always show others your best side. You hate the titled rich but only because you are not one of them. And if you do choose to swear avoid saying fuckin' and pronounce the word the way it is correctly spelled. It is fucking.'

Bill's temporary existence in Canada started to look good. Conflict elsewhere tore people apart. The suffering back home was enormous but there in Saskatoon Bill's light of reason had been muted at behest of his appetite for pleasure.

When Bill was able to leave Base he spent time with Sophie, and they became lovers.

Their kissing had become more urgent and desperate with their relationship and she'd responded wonderfully towards him each time they met.

Sophie was frequently upon him as she sought to release her passion. Her tongue darted anxiously inside his mouth as she felt him pressed up hard against her. She seemed to want him more than she'd ever wanted anyone. They kissed until their lips grew sore but were unwilling to stop and let each other go. Each time Sophie's skirt fell away, Bill's fingers fumbled with her blouse, then down to her buttocks as she stepped out of her panties.

Her ardour continued to rise to a crescendo. Then Sophie would take Bill inside her in one great shuddering gasp and often howl with pleasure. He wanted to stay forever, his head pillowed between her breasts, as he savoured every delicious moment before reloading for another assault. As soon as he was hard she grabbed him and shoved it in. The harder he pressed, the more she took. He thrust as if she couldn't get enough. Mostly a race to another photo finish as time and time again, he died inside her.

'No man has ever made me call out before, Bill,' Sophie purred.

While Bill was with Sophie, Richard invariably went off with someone, sometimes Anne. The only loose end was Olly. One night he knocked rather sheepishly on Anne's door with a bottle of champagne and two long stemmed glasses in his hand, and was ecstatic when she didn't send him away.

Near the completion of each novice pilot's training plan, they were assigned a complex solo task of navigation. Given a route and map readings, they were sent off on their own to complete a cross-country round trip of a few hundred miles without assistance.

After another heavy night with Sophie, Bill was given his flight instructions and decided there must be an easier way to achieve this incredibly complex and silly task.

He adjusted these unnecessary and befuddling instructions a bit, and created his own simple formula.

His brilliantly conceived plan was to side-step these stupid map co-ordinates and instead follow the river as his continual landmark. Bill knew the basic laws of physics. Water runs down hill. If he followed his ribbon of blue, easily visible from the air, he would save himself considerable time and effort. If there was any hint of doubt, this same river was conveniently located on his map in the handy form of a distinct blue-ish line.

As they departed for their designated flights Bill waved goodbye to Richard and Olly who were both to take off ahead of him.

'See you back here in time for dinner, Bill. If you get lost follow me,' Richard said, and with a generous wave of his gauntleted hand he was gone.

Olly soon followed, then Bill.

In what could only be summarised as some sort of a misguided daydream he headed northeast and followed his beloved river. He set out with unbridled enthusiasm, of not one but half-a-dozen eight-footer Hun troopers. And in his pocket was a photo of Sophie. So convinced was Bill how much better his idea would work out than what he'd been instructed to do, he didn't bother with a back-up plan. And why should he? Plan 'A' was perfect. He was the man and nothing could go wrong. He had a date with Sophie for the weekend and needed sleep to recharge his batteries before then.

It was a truly magnificent and conquered day even before it had started. Up there the air was pure and clean. Through the gentle cruising speed of about 90 mph he marvelled at crystal clear blue skies, although there was hint of a slight haze. Sharpness of his guiding river below was clearly visible from his cockpit. Secured by the additional comfort of his indelible printed blue line on his map. Bill became at peace with the world. He climbed to a ceiling of 10,000 feet and decided to stay there. With weather close to perfect, only a small amount of high cloud was visible above him at 25,000 feet or more.

War, what war? His mind replayed his last sexual encounter with Sophie. She'd been so willing, so accommodating. Infatuated with her, it was if he'd found his *Helen of Troy*.

After he'd been flying for some time, he noticed the slightest hint of possible concern, in the distinct lack of his river beneath him. A bank of cloud had come between him and his river and the best of the day was behind him.

Time dragged slowly as he flew lower and lower over the ground in what looked deceptively white, as in—snow cover.

Despite its transformative enchantment of ordinary surfaces his disquiet changed to alarm.

Oh, fuck! Something must be wrong with his original plan. Rivers don't disappear and he shouldn't be so far north.

Possibly he'd failed to notice when the river trickled to a stream and meandered away or changed direction. He was unable to sort out his compass bearings or understand the co-ordinates.

Toto, I've a feeling we're not in Kansas anymore.

Bill flew on in a gathering state of panic. A sour taste came into his mouth. He was lost. And as the sun prepared to sink below the horizon, darkness gathered around him. He found himself in a mauve haziness and became frightened. Visions of dying during a botched training exercise floated across his mind. What would Lily and Fred say back home when they received their dreaded telegram?

His mum, he knew would shed tears, and together they would try to sound supportive. Fred was proud of his only son learning to fly to become a pilot. Others might not see it the same way.

'Bill, stupid prick, didn't even make it into combat,' some would say.

'He got lost in Canada. Ran out of fuel on a training exercise and crashed. What a dickhead he was,' others would snigger.

Glorious though he had felt up there alone Bill knew he needed to extract his digit and get back to reality. He had a brainwave. It was customary for towns to have their names painted in large letters on their town silo. As this vast land was not yet a war zone, luckily Canadians had not discontinued the practice. Memories of near Base flashed into his mind, *S'toon*. If he came down low enough and flew around a wheat silo he might be able to see where he was.

That's a bloody good idea, Toto.

He slowed to less than 80 mph and circled the next silo he saw as tightly as he could fly. It would have caused some considerable concern to any locals out enjoying the air. A solitary plane, flying in ever decreasing circles around and around their lone wheat silo. Unbeknown to them the pilot was attempting to read the name of their town from reflective beam of his landing lights. After about three or four sweeps, he made out the name.

Now, his torch held in his mouth, his map spread in folds on his lap he knew the name of the town and he could find where he was. He was no-where near where he called home. He was past his SPNR (safe point of no return).

Irrespective of the scale on his map, one inch meant a long way and was two- folds off his course. With the aid of his compass Bill proceeded to fly southwest in an attempt to find his way home.

Did he know where he was? Yes, of course! No! Bloody liar!

Bill had never had much idea of direction and now in total darkness, the option of him waiting around for the sun to rise, held little appeal. Knowledge of how the sun rises in the east and sets in the west is of poor value at night. He looked up at a myriad of stars through some cloud but they seemed dim and far away. Irrespective they all looked similar to Bill.

Astrology hadn't helped the rabbit running into the field.

Stars swam in dizzy patterns. Planets as big as footballs swung ominously about. He felt guilty abusing astronomy but consoled himself there had always been something dubious in his relationship with nature. Best option at hand was to focus on a single brightest star and fly towards it.

Dark as it was, it was only going to become darker. He pushed unsavoury thoughts of fuel tanks run dry from his head, as he desperately stayed focused on that one damn star. Sheen on his forehead was the sweat from tension and fear.

Low on fuel, and with difficulty thinking of a good excuse, he found what looked like his home Base aerodrome he'd left many hours earlier. But with lights out, most likely he'd been given up as another lost pilot.

He buzzed low over the barracks, his fuel gauge wrong side of *Empty* and created sufficient commotion to alert someone, anyone, to take an interest in him. With hastily switched on runway lights he could at least see to land. After he touched down the lights went swiftly off again.

With about half a pint of petrol sloshing about in the pit of his fuel tank, Bill was convinced this was a test he'd failed.

'I have not failed. I've just found 10,000 ways that don't work,' Thomas Alva Edison (1847-1931)

Chapter 14

RICHARD

No one was overly concerned at Base that Bill had returned from his excursion late. It seemed they always expected the unexpected. He'd departed as planned, signed out his plane and returned to sign back in. That was acknowledged as achievement enough. Bill shrugged. Okay, he thought. There and back again is justification for a pass.

'Where did you end up, Bill?' asked FOB.

'I'm not quite sure, Sir,' answered Bill warily.

'Oh well, at least you're honest. That will do for me.'

A little bit flimsy, thought Bill.

'Aye, but great men are naught always wise,' Olly offered shrewdly.

As night progressed, Olly turned in. Those who remained became slightly pop eyed. Others were concerned how one plane had not returned and was to be reported missing. It had become a big night.

Bill's instructor FOB was delighted Bill had used his initiative and found a way to resolve his problem.

'You did in effect get yourself un-lost, Bill. Not unusual for a pilot to become disorientated and get himself well and truly misplaced. Well done.'

Bill now understood basic advantages of map reading, how to use co-ordinates and what it felt like to be lost and in deep do-do. He'd learned a valuable lesson and in the process, fear had rubbed away a great deal of his effusiveness. But he'd survived and passed their test.

No longer did Bill feel dumber than a headless chicken. Now he wanted to share his moment of triumph with Richard. He went to find him with newfound enthusiasm.

'Where is our aristocratic teacher, Richard?' asked Bill.

The steward frowned.

'Haven't you heard, Bill? Richard, his is the missing plane...'

The words were harsh and unpleasant. Like the sound of chalk scraped on a blackboard. Bill felt pole axed. He couldn't believe what he heard. Waves of panic swept over him.

His vision blurred like an after-image, tears swam in his eyes. He blinked as he tried to speak. Words didn't quite form as intended. That awful tinny taste he'd had in his mouth a few hours previously when he was lost and felt doomed returned. This time it sickened him to his stomach. His hands washed together, his face a mask of concern.

The steward seemed to understand. 'Don't take it too hard, Son. You'll get used to that. Tomorrow it could be you!'

'Bloody encouraging. Thanks!'

Then the steward pushed a large measure of whisky in front of Bill. 'Here! Have a nightcap. This one's on the house.'

Alcohol washed into his mouth and he savoured the bitter heat into his belly. Then he felt nauseas and wished he hadn't.

At the exit, he collided with the doorframe and felt conspicuous. When he turned to see if anyone had noticed, it was as if the world was staring at him. He knew he had to get out, they must not see what could be taken as a sign of weakness.

Bill had seen others react badly, even cry of sadness in the slums. Over loss of a pet or a toy but no good ever came of it. A tear was not easily drawn from him but outside was where he wanted to be. There he could compose himself in private.

Instead of going to find Olly he took a walk to regain his composure. Bill's thoughts were of the young aristocrat from a world of chimneys. A man who had befriended him, hadn't returned home, and was probably dead.

Later as he lay in bed, he came to realise this situation was new to him. Bound to happen sometime but he hadn't expected it now and certainly not Richard.

As he dozed Bill couldn't help but think about Richard's tattoo advice. 'From a practical, if not dubious point of view, old boy, why would you want to unnecessarily stamp yourself with such clear and indelible identification? What if you needed to disappear or wanted to change your identity?' Richard had said.

Thoughts that Richard might not be dead at all but alive helped Bill through that night.

When Bill broached the subject, Olly was dubious.

'Aye, but most likely he's crashed, Bill. Summat's happened to him and naught good, by gum.'

'But he seemed so sure of himself. Before he took off he even said, 'See you back here in time for dinner, Bill. If you get lost follow me.'

Olly shook his head in sadness. He'd liked and admired Richard.

'Aye, it's hard to accept. Summat of a shock to think our show pony of a home grown aristocrat's naught coming home.'

'The image of Richard's wave before he took off is etched into my brain, Olly.'

Anne took the news better than Sophie who had hysterics even though she'd barely known Richard.

'It's a bloody shame for sure,' said Anne. But although she'd been intimate with Richard she was familiar with death. She saw more at the hospital than Sophie ever did. She was used to knowing people and liking them, only to hold their hand as they died. She had become hard in a way few wanted to.

In their Mess a lone glass of whisky sat on the bar, where Richard had often stood. No one ever asked what happened to its contents. Bill raised his glass. 'Cheers, Richard.'

He waited until he was in command of himself and left. Then he blew his nose into his handkerchief, very, very hard.

Chapter 15

LICENCE TO FLY 1943

Bill's letters from home told of most down with the flu and the old country in chaos.

'They say, when they can get it, there's a new product on the market called SPAM.'

'Aye, I've heard of that,' laughed Olly. 'It's summat posing as meat.'

'On the brighter side comrades, Rodgers and Hammerstein's musical Oklahoma has had rage reviews on Broadway,' offered Duncan.

War wise Barnes Wallis had designed his famous bouncing bomb and Wing Commander Guy Gibson had delivered them on target to burst two dams. Guy used a new method to guide Lancs onto their target. VHF radio transmissions. Unbeknown to Bill at the time, the RAF was excited at Guy's success and developed his strategy, but in the months to come the Master Bomber technique, as it would be known, would have disastrous consequences.

But Bill had other things on his mind. Prior to his graduation was a qualifying flight with an instructor who had a reputation for being difficult to get on with. Olly flopped down onto the sofa. Swung his legs up over the end arm and leaned back. 'Aye, he's a naught but a total bastard. Uses structure to control.'

'Forget it, Olly. To get this far and fail because some miserable arsehole enjoys playing games at our expense. It's too appalling to consider.'

'Aye, but this particular instructor takes great pleasure doling out naught but ridiculing inexperienced pilots. He's unpopular even with his own colleagues.'

'One cadet told him he wanted to fly naught but Spitfires. There's a long way to go before that, Cadet,' mimicked Olly, 'you might not make it for a number of reasons. You seem a bit brash. That can cause accidents, which can cost us aeroplanes and they cost money, lots of money.'

'Yes, Sir, but I'm sure I'll be able to cope, Sir,' had pressed the cadet.

'Yes, I'm sure. But it's not about you. There are plenty like you for us to choose from, Cadet.'

There was an air of worry and concern on every cadet's face to get this far and fail.

After Olly had been up for his turn, he gave Bill some advice.

'Aye, you watch out, Bill. Bastard waits until he thinks you're naught paying attention and then he turns bloody engine off to see how you react to an emergency.'

Bill was thoughtful. He knew the procedure. Lower the nose of the aircraft to gain momentum without the big fan up front forcing you along. Lay the plane over and restart the engine. Not unlike the procedure to bump start a car in gear downhill. Provided the ignition is on, the engine should restart.

'Ignoring elements of luck, Olly, two principles for keeping any plane airborne are always 'money' and 'wind speed,' laughed Bill.

'Aye, they are, Bill. Then son-of-a-bitch shouts summat at you, or grunts. But he was all disapproval with me.'

'You mean he failed you?'

'I don't know yet, Bill. He's adding up his scores, but I'm naught confident. Aye, I felt a right proper dick-head when he reached forward and turned engine back on.'

'So his only purpose is to establish your reaction to an emergency, not create one?'

'Aye, when smart arse switches engine back on, you can level out, and look forward to summat like changing your underpants quick smart.'

Armed with this privileged advice Bill felt more confident. He was ready for his turn.

Up they went to 10,000 feet, levelled off and did a spin to the left. He recovered and backed up to 12,000 feet and did some aerobatics. Then around and around they flew. Everything asked of him, Bill did without any trouble. He stalled and recovered well.

Bill watched the instructor out the corner of his eye. *Go on then, you bastard. I'm ready for you. Turn the fucking thing off and we can both call it a day.*

But nothing out of the ordinary happened. After a while Bill stopped concentrating on the instructor and began to enjoy the sheer bliss of flying. The air was pure and clean. What a shame to have to fight a war.

Suddenly the engine coughed and stopped.

You crafty old bastard, I never saw you do that, thought Bill.

Quick as a flash Bill lowered the nose and executed his preconceived dive.

Copybook stuff. He'd been prepared and waiting to do it. They continued to dive.

Nearer and nearer to the ground they flew as they fell from the sky. Bill lost altitude fast in a near vertical dive. When will he turn that bloody engine back on? He turned and looked squarely at the instructor. He was as white as a corpse. Hanging on grimly to his seat in their state of freefall he had tell tale signs of vomit down the front of his otherwise spotlessly clean tunic.

The penny dropped. This was not a drill!

Oh, fuck! Their plane was in a controlled dive all right but death was within arm's reach. The altimeter spun as quickly with their decrease in height as the speedometer was spinning their gain in speed. As their speed increased to an awful whine the entire plane began to shake as if about to disintegrate.

Bill had been trained to react in an emergency. Don't panic! He panicked. Then his training kicked in. He passed rapidly through the procedures to restart the plane. Only seconds, but they seemed an eternity as the ground rushed up to meet them.

Oh, fuck! I can see branches on trees...

The engine turned over once, coughed, spluttered and as the Gods smiled down, roared into life. Bill couldn't feel the muscles in his cheeks due to the G—force and his vision started to blur. The sensation, as if all air was sucked from his lungs and his blood stopped flowing as he pulled back on the stick with every fibre of self-control he could muster overwhelmed him.

All he could think was; No! I don't want to die!

He had to control the lift. He coaxed. Too abrupt and his engine might stall again, leaving him no room for a second chance. Gently he forced the stick up. It felt like an hour or so. One of those moments when time seemed to slow to a stop but gradually the nose began to level out. Vibration ceased. Noise pitch changed. Settled into what could only be described—as the sound of 'a thousand hummingbirds resting on the hand of God.' Then he pointed them upwards again.

'It made a sweet, sweet sound, Olly,' Bill said later.

'Aye, just like God to take a special interest in that one, Bill.'

But this was where FOB's training paid off. Know it, follow it, do nothing else.

Being scared had changed to angry. His earlier vicious and violent responses were now replaced by a greater determination to succeed.

He circled low around the aerodrome once, a bit like a lap of honour as he prepared to land. He considered a victory roll or two, but with a look at the state of the instructor, decided not to push his luck. He landed after one ungracious bounce and elated, he taxied back to the apron.

The instructor could hardly stand. Drained and visibly aged, he had to be helped by ground crew from the plane but no one wanted to touch him.

Bill shuddered. If he must handle vomit, at least let it be his own. But he remembered thinking to himself, how the instructor must have taken seriously the nonsensical advice to eat carrots. Air Ministry always insisted pilots ate carrots by the bucket-loads to see in the dark.

That or the instructor liked them a lot.

As ground crew helped the instructor away, his clipboard clasped firmly under his arm, he looked to Bill like a retired greyhound clutching a stuffed rabbit.

Bill felt good, until he saw Richard. But no it couldn't be. Adrenalin must have played tricks on his subconscious mind. He knew Richard wasn't there but would have sworn he saw him give his thumbs up in recognition of Bill's efforts. Bill didn't believe dead people could return but he sensed something. It felt like the brush of wingtips against his shoulder. Bill turned then twisted back. Richard was gone—but his image had seemed so real.

He wanted to tell everyone he'd seen Richard and he was all right but he suspected if he did, they might never let him fly again.

'Aye, best keep that gem to yourself, Bill. Tell them naught about that.'

Bill never saw Richard again.

'These are good days for me, Olly. Every day is a fight I can win.'

Later in their Mess the instructor explained.

'I've seen that reaction before, Bill. I felt sure we'd survive. When I saw you staring down at the ground, shouting not in fear, but anger, 'Come on, you bastard, lift. You fucking cow, lift. Lift, and fly you bastard. Lift, lift and lift! That kind of response is usually from pilots who know they're in trouble but are intent on getting out of it.'

Bill remembered nothing of his outburst and was embarrassed. He felt a total twat and it showed in his facial expression. The instructor read Bill's thoughts.

'Not at all, Bill. Christ, that's better than crying or giving up. When I saw that in you, I was sure we'd get out of it. Remember nothing kills a pilot quicker than panic.'

'But why didn't you help me?' asked Bill.

'If I'd helped you, I would have had to fail you. We didn't want that to happen, did we? What a waste after you completed 180 flying hours.'

On 26 November 1943 there was no ceremony, no passing out parade.

An announcement of training results appeared on their lecture room notice board, unheralded. The wait had felt like a lifetime. The list of hopefuls had thinned but he, Olly and Duncan's names were there. Others had been reassigned.

Bill was handed his cloth pilot's wings and Sophie sewed them on for him. He was a fully-fledged solo pilot and could soar like an eagle. For Bill it looked like a step in the geographical direction of his dreams.

The next two weeks were spent flying larger twin-engine aircraft in duo. They stalled, they dived and they recovered. They did spins. Nothing was attempted at less than 8,000 feet. Bill added a friendly margin and decided in future, if there were to be a future, 10,000 feet minimum would do him nicely.

What he didn't know yet was what the RAF had in mind for him. Captain Hawkins might have been right. Perhaps they didn't know themselves.

Richard was listed as KIA—Killed in Action. The acronym used.

His plane, or what was left of the charred remains, was found in a field beside a main road.

'Aye, summat strange about our missing aristocrat, Bill. Naught of his silver neck chain nor that fuckin' great medallion, neither.'

Bill was thoughtful. 'It's bloody strange. You'd think they'd find something of him.'

'Aye, habeas corpus Bill.'

'Corpus in absentia,' laughed Duncan.

When they asked around of those in the know they received a quizzical shaking of heads.

'Not surprising really. It was difficult enough task to identify bits of plane, let alone a pilot,' said FOB.

Richard's fate might have been a subject of some conjecture, but not for long.

There were many variations of what might have occurred with countless numbers of young men to join Richard's endless list.

One expert stated the wreck of Richard's plane appeared to have exploded on impact with the ground. The intensity of the fire aided by the fuel tank was the equivalent of a cremation.

'Stranger things have happened,' they said.

The group were told they would receive their final 'orders' after their return home.

'Bill and Olly, you should expect to be assigned to a fighter squadron. Most likely Biggin Hill. Perhaps it will not include Duncan though. He has a question mark above his head when it comes to his reflexes,' confided FOB.

Duncan had not fared well in the orientation tests. Yes, he would fly but perhaps slower planes, as in bombers. Duncan was disappointed. He'd set his heart on defending the home skies as a fighter pilot.

'My family know people,' ranted Duncan. 'They're people in high places. We'll see about this crap.'

FOB, along with Bill's other instructors, reported his quick responses were excellent. Also his level headedness in suicide dive situations and his extraordinarily good eyesight particularly at night gave him a natural bent. He was, they believed, an obvious choice to fly faster and more manoeuvrable Spitfires or Hurricanes.

'You're a natural, Bill. If ever I've seen one,' FOB said. 'Those ninety-degree dives don't agree with everyone. And you never vomit when you're the pilot, only when you're not at the controls.'

The test FOB referred to was something each pilot had to contend with. Up as a passenger they did one circuit but it comprised two ninety-degree dives in quick succession before landing. The passenger then exited the plane and stood to attention in front of the examiner. In all for about three minutes, his reaction being closely monitored.

'Aye, passing out or naught, even a tendency to faint and you're out,' said Olly.

'Even any rolling of eyes, is a fail,' groaned Duncan.

'Aye, if they continue to roll after about ten seconds or so, summat happens all right,' laughed Olly. 'Duncan's now scrapped from naught but flying bombers.'

Olly scraped through, but Bill's eyes never spun once during the test. True, he'd thrown his heart up but in the plane, before exiting. He was overjoyed with his results.

'I expected to pass, Olly. I had a Viking sense of entitlement, but never in my wildest dreams did I expect accolades.'

They spent the next three weeks flying an assortment of planes, before their scheduled return to England. Olly fished out a small gift-wrapped parcel from his pocket and handed it to Bill. As he did, he looked embarrassed.

'What's this?' asked Bill.

'Aye, it's just a little summat. Naught but a small memento of our time here in Canada,' beamed Olly, as he shuffled his feet awkwardly.

Bill was speechless. 'You shouldn't have,' he mumbled. 'I didn't get you anything.'

'Aye, but it's naught. I wanted to thank you, Bill. You were the only one who visited me in hospital. Even Richard never did that. You've been a good friend, Bill, and I thank you.'

Bill fumbled and unwrapped a small thermometer. It had as a background a seal and polar bear.

Olly turned on a cheeky grin. 'Aye, a reminder of temperatures rising in Canada.'

'Soaring might be more appropriate.'

'Aye, I couldn't get one with a picture of an Inuit girl, nurses, or mosquitoes, by gum.'

They laughed. Bill thanked Olly and slipped the memento into his pocket.

He felt confident. 'All we have to do now is survive our voyage home without being torpedoed or hit by a tsunami.'

'Aye, naught to worry about. We'll take our places as fighter pilots soon enough.'

Bill had enjoyed the camaraderie shared in Canada, but he'd come to a number of conclusions. He liked women better than men. He had nothing against men but in the RAF there were too many of them. Thousands jostled, shouted, swore and fought. Some like Olly and Richard had become Bill's friends, but women he'd decided were softer, gentler, cleaner, sweeter smelling, and altogether better company on a one to one basis.

To find someone he liked and respected, someone he could love and who would love him in return, would be his quest. But after the war. Best get that out of the way, first.

He hoped he might have found what he was looking for in Sophie, but it wasn't to be. They'd met up in town, a spontaneous arrangement. In her haste she'd forgotten to remove her wedding ring.

'What's with the ring, Sophie?' asked Bill.

Sophie was devastated. She had to admit she was married.

'I'm sorry, Bill.'

'Why are you sorry, Sophie, because you're married or forgot to remove your ring?'

Had Bill realised her young husband was away at war overseas, he would never have allowed himself to become involved with her.

'It's not me you should be apologising to, Sophie. It's your poor bloody husband risking his life to come back to you. You carry on like a single girl and it's not right. Had I known you were married I would never have come anywhere near you.'

There were tears, and more tears, Sophie begged Bill not to leave. When she realised he was adamant she wanted him to at least keep in touch with her.

'Aye, but perhaps she was interested in you, in case summat happens to her hubby. What if he naught makes it back or summat?' suggested Olly.

It was unacceptable to Bill. He felt like a delinquent and couldn't get away from Sophie fast enough.

'I can't see past her dishonesty and lies. They would pervade any future relationship between us.'

Olly shuffled his feet and studied the ground. 'Aye, you're right, but for the sake of great sex your relationship could be summat of a drug remedy cycle,' he sniggered.

'Sex isn't everything. If she's quick to be unfaithful to her husband, what happens when she tires of me?'

Bill took solace in his pipe. Thoughts of Sophie he kept to himself. He remained in a less than a rosy glow.

'This Sophie you were seeing, Bill. If it's over between you, would you mind if I call her?' asked Duncan.

'It wasn't so much a relationship as a fling, Duncan. Go knock yourself out. Just don't expect to play with me in my sand-box.'

'Aye, Duncan loves Jesus, cleavage and arse, by gum,' laughed Olly.

There were a couple of times when Bill and Olly saw Sophie and Duncan together.

'Duncan might not be leaving much for her old man,' Olly said wistfully.

Bill decided she no longer missed him, and she'd replaced her husband with another, but he realised too late he longed for her. To see her again had awakened his feelings for her. He'd never carefully considered the realities of his situation. Nor had he asked after their quarrel if she wanted him back. He knew he desired her but he'd told himself a lie.

'If life had continued as it had in Canada for much longer, Olly, I would have died of over indulgence.'

'Aye, but what a way to go, by gum.'

Another surprise for Olly and Bill before they left Canada was the discovery of the near naked body of a once attractive nurse named Anne, wearing a long platinum blonde wig. It had been dumped beside the road north of town, coincidentally not far from the site of Richard's crashed plane. Canadian Police investigated this incredibly brutal murder possibly as a crime of passion. Anne had been raped and beaten to death with a blunt object.

Neither Olly nor Bill had ever known any more about Anne than her first name and her work at the hospital. But it was definitely the same girl who worked as a nurse at the hospital full-time and attended church hall dances to further hospitality with RAF cadets.

Bill never learned the outcome of the police investigations, or whether anyone was charged. What he heard was news from home. Doom had cast shadows over the RAF and in particular Bomber Command. Their ranks of young aircrew were being sadly depleted. It was a bitter period and losses were high.

'Thank Christ we're going to fly fighters,' Bill said.

He, Olly and Duncan bade their final goodbyes to Base, the town of Saskatoon, and the ghosts young men sometimes leave behind. Bill tried hard to overcome his naive expectations of a married woman but found it difficult to erase Sophie from his mind.

He thanked FOB for all his hard work, which wasn't a chore as he'd come to respect and like the man.

'Do us proud, Bill,' said FOB.

'I will, Sir.' Bill snapped to attention with his smartest salute.

Both men knew they would never see each other again.

Soon after Bill, Olly and Duncan were on a ship back to England.

They arrived 21 December 1943 in time for Christmas. Russian armies might have pressed on and captured Kosovo, but Bill's heart was in Canada.

'Only two things are infinite, the universe and human stupidity, and I'm not sure about the former'... Albert Einstein (1879 to 1955)

Chapter 16

BOMBER BOYS JANUARY 1944

Returning from the wide-open spaces to the overcrowded land of their birth and its unpredictable weather came as a shock to Bill and Olly. It reminded Bill of when he'd made his way to Lord's Cricket Ground to join up one year earlier. The weather was identical although he realised he no longer dealt with the same consequences of defeat. His determination to improve himself had paid dividends. Nonetheless, disappointments in losses such as Sophie, Anne and Richard were hard to put behind him.

Now, as if outcasts to an abandoned people, their cloth wings firmly ensconced on their battledress uniforms they moved slowly towards their deliverance. They also received their long awaited Sergeant's stripes with increased pay.

'Aye, and now we're back in the old country, I might need to lose my new sunnies.'

'Not yet, Olly, they might be handy flying above the clouds in daylight.'

Their next surprise was being assigned as pilots to the Bomber Boys.

Bill couldn't contain his disappointment. Only Duncan went to Biggin Hill, the best-known Spitfire aerodrome in Britain. Olly and Bill were to fly Lancasters, primarily night bombers. * 5th Footnote

'So much for flying in daylight,' moaned Olly. Bill was speechless with dismay.

Now he had a further disappointment to add to his list.

Amid squeals of tyres and grinding of gears, trucks deposited them and a myriad of other fresh faced young aircrew to guardrooms all over England. Bill and Olly were assigned to an aerodrome north of London enveloped in a dull canopy of grey. As they clambered out of the transport the wind hit them. It was bitterly cold.

They found the ground crusted with ice, a concrete runway, conglomeration of hangers, makeshift buildings and a host of hastily assembled Nissen huts.

'Aye, look at it, Bill. Draughty corrugated iron with naught but less than sanitary conditions on summat of a hastily constructed wartime airfield,' groaned Olly.

He already missed the luxuries of Canada.

A single road from Base to the bomb dump meandered its way past various buildings. Beneath fresh snow recent rains had turned the gravel surface into a sea of mud before it became uneven icy concrete brought on by frozen conditions.

The bomb dump comprised a depression in the ground where anything could be jettisoned from unexploded bombs to discarded aircraft parts. Elsewhere lower grade ground crew rugged up to their eyeballs slowly swept snow from the runway while the squadron waited for clouds to lift and freezing conditions to ease.

Nearby were villages. Some had quaint sounding names but each sported a colourfully named tavern with a welcome log fire. No doubt their publicans were thrilled at the prospect of additional customers slap-bang on their doorstep. Visions of hundreds of thirsty swilling airmen as a captive audience would make their cash registers sing with joy.

After Bill and Olly had settled in the CO elect summoned them. Not yet thirty years old he was already a Group Captain. As bald as an apple, his face flushed red. Although a temporary appointment, he was intent on making an impact.

'This squadron's assigned to high-level night bombing of military targets; factories producing armaments, fuel and oil depots and trains carrying munitions. Do your jobs and bring my planes back intact and we'll get along,' he said without a smile.

'Yes, Sir.'

Outside Bill said to Olly, 'who the fuck does he think he is when he's at home?'

'Aye, you're right, Bill. He needs dragging off that high horse he's sat on. Not only weather's bitterly cold here.'

In their Mess they met other ops crews, which included a tired young pilot nick-named Tache. It was a fitting synonym as he was one of a few who wore a distinctive trimmed moustache.

'Aye, CO's naught much of a friendly chap,' observed Olly.

'Welcome to the losers' lounge. He's a nerd and several sandwiches short of a picnic,' said Tache. 'He's more interested in graphs than lives. Doesn't give a rat's arse for us. Never has. He's always on at us to edge back the throttles to lessen power and conserve fuel. He only cares about his wretched planes. What he is, is an ambitious little turd.'

Bill nodded. 'He made that clear.'

'Aye, but won't that mean flying lower?' queried Olly.

'Certainly does, but he won't hear about that. Thank Christ he won't be here long although heaven knows who might replace him. And another thing, don't trust your fuel gauges! They're not reliable. It's best to dip your tanks to ensure you've got enough fuel. Even the best of ground crew make mistakes. And watch out for careless ones because there are plenty of them, and their fiddles on the side.'

Bill was not happy. Not only had the severe frosts and snow turned the mud hard as a rock underfoot, now this crap. Dumfounded the RAF had completely ignored FOB's recommendations; how his skills were perfectly matched to faster more manoeuvrable Spitfires and Hurricanes, Bill put in an application for transfer.

That really upset the CO to be.

'Aye, not a great start. Prick has naught good in store for us, that's for sure.'

Bill's request to be reassigned in accordance with his instructor's recommendations was denied. A decision, basis of which Bill never understood, but assumed like others, the powers knew best.

'They know fuck all,' said Tache. 'Trouble is they won't bloody admit it.'

'Aye, but how stupid might they really be?' asked Olly. 'What if they know naught about this nor running any war but it's our lives on the line?'

'They really don't have a bloody clue, which goes a long way towards explaining why it's taking so long to win this war.'

On the surface the place appeared to run like a well-oiled machine. Everyone looked busy.

Tache told a different story; 'It's quite the opposite, chaps. Disorganised chaos! Bomber Command doesn't give a toss for anyone other than themselves. They call on crews to perform one desperate op after another. Bloody losses here last month were as high as sixty per cent. You survive, if you're lucky, day by day.'

Their Lancaster familiarity training comprised only a few hours. There was no time for a structured conversion course. They were told to sit in their cockpits to learn the layouts. Following fleeting introductions by someone on ground staff who called them, 'sir,' they were presented with handbooks. After about an hour spent with other operational pilots they were told to test themselves. This Bill did by closing his eyes and running his hands across the instruments while naming them to Olly. He read the manual by going straight to the parts marked with asterisks. He noted stall speeds, revs and fuel consumptions. Bill and Olly were deemed ready to fly Lancasters.

'Aye, they look summat past their prime to me. Yet they're not old.'

'They're clapped out already,' said Tache. 'You'll get used to it. It's not only the likelihood of death keeps you awake at night, but also the method of dying. Take your choice; flak or being shot down by a fighter. That or fall to earth without a parachute from a disintegrating aircraft.'

'You paint naught friendly pictures. Why are we talking about death?'

'Why not? I'm being honest. You won't find much of that outside of our own crews here. All the brass ever wants to do is string us lies, do bugger-all and get promoted. It's probable none of us will get out of here alive.'

After the adrenalin packed smaller planes and Bill's dreams of flying Spitfires or Hurricanes, Lancasters were altogether different. They looked bigger than a football field and behaved dopier than an elephant on medication.

Bill met his assigned crew of six. He wanted to work them out, test how far he might rely on them, and find out their weaknesses.

Joe, his Flight Engineer, was also his co-pilot. From ground crew with five hours of learning to fly on a Tiger Moth to his credit, he was enthusiastic. Lancasters being a four-engine bomber meant quadrupling most fuel and electrical systems.

Co-ordinating these was a tricky job for Joe who knew more than most about packing fuel logs and tapping fuel gauges as he anxiously eyed oil temperatures. What he didn't know he would learn quickly during flight as their ragged armada thundered on. Joe also assisted Bill when he took over as pilot for any basic straight and level flying.

Navigator Ben turned out to be a canny Scot. He was a smaller sturdy man as many Scots are, but as hard as, with piercing eyes set in a sad face. Sole survivor of an earlier crash landing, he was not chatty and headed towards freakier pathological tendencies. His plane had had a 'hang-up,' a term given to a bomb that failed to release. Being a more thinly cased blockbuster bomb, it was often called a 'cookie' but at four tons, it was a highly sensitive weapon. Packed with high explosive TNT it had been loaded with more than a little trepidation into pride of place in the middle of the huge bomb bay. In appearance it resembled nothing special. Two 44-gallon drums bonded end-to-end. No one was surprised when it exploded as the Lanc touched down.

Thrown clear from the wreckage in a rare combination of good fortune, Ben had listened to screams of his fellow crewmen, as they burned alive. The only saving grace was the horror was short-lived but long enough for Ben to witness the shrinking of the bodies of his friends; pilot and flight engineer. He'd watched them through the bubbling Perspex. Their bodies blistered up then shrivelled in the intensity of the heat.

Ben was given a tot of rum by the medics and a compassionate leave pass for twenty-four hours. The powers would never know how close he came to going AWOL.

Ben shuffled his charts as their Lanc swung ever deeper into the abyss of total darkness. He worked through the process hindered by his own demons, but assisted by consuming copious quantities of alcohol daily.

Unfortunately booze didn't help. It compounded his guilt and rolled straight off him.

'Trouble is the RAF isn't consistent. Aye, they don't play by their own wee rules when it suits them,' complained Ben. 'After I changed squadrons and got this new wee crew, I lost my accrued ops. It's as if I'm back at the commencement of my tour, lads, and I'm none too bloody happy about it.'

'If they keep doing naught but that, no one will ever retire,' observed Olly.

'They'll do it all right and they'll stamp you LMF given half a chance,' said Joe.

Tache agreed. 'That's right. Because it's an easy out. They prefer to stamp a man with 'Lack of Moral Fibre' because there's no comeback on them.'

'He's right. Its stigma is fierce enough to eclipse fear of death itself,' said Joe.

Olly agreed. 'Aye, it's also difficult, if naught impossible to defend.'

Bill was thoughtful. 'Trouble is, their doctrine of discipline is to handle any dissension immediately,' he said.

'Aye, that's why morale is naught but at an all time low,' agreed Ben.

'A pilot I knew with a wife and a small child, got himself stamped LMF by his CO because after thirty-three ops he objected, to more,' said Tache.

Olly became angry. 'We're only supposed to do thirty! What the fuck do they expect?'

Tache ignored the question but continued. 'On one parade ground not far from here their unit recorded eighty aircrew beginning a disciplinary course, of which twenty-five were officers, and twenty-one were NCO's. And another thirty were court martialled at Norton for being AWOL. All the RAF knows is the big stick. I tell you, they don't put any of that in the *Daily Express*.'

They shuffled their feet and began to feel the hopelessness of it all. Work brought few rewards besides survival.

'Perhaps, Friedrich's quotation is true?'

'What quotation?'

'What doesn't kill you, makes you stronger.'

Olly frowned. 'There's naught about that in their bloody brochures either.'

Wireless operator Mack was Canadian and a larrikin. If it hadn't been for him finding solace in the form of a bottle his face would have been as clean and straightforward as a peeled apple. He already had the ruddy, purple hue of a drinker. Micro-fine spider veins rambled across his nose; the look of the alcoholic. He was a big-hearted man with broad shoulders and a grin to match but today his whole face glowed crimson.

Mack held little respect for authority as he sat in front of a big Marconi wireless set with its distinctive coloured tuning wheels. He operated the flare pistol in emergencies, Morse key, looked after his code flimsies as if they were maps to buried treasure and hoped like hell he'd never have to eat the edible rice paper code book, thoughtfully replaced by the powers each trip. He also handled an array of radar and night fighter warning equipment. His seat was an uncomfortable piece of timber suspended on poles in the metal floor.

Bomb-aimer Sandy was a serious man who collected target maps and in between was an alcoholic due to peering fretfully into the darkness for the uncertain glimmer of a navigation beacon. His biggest fear was to see for the first and last time a shell coming at them from an angled upward-firing cannon on an enemy fighter. Few who witnessed lived to tell the tale afterwards. It was an ingenious German weapon believed to be the most diabolical of all. *Schrage Musik.* In combination with advanced Lichtenstein hunting radar, it was unannounced slaughter without tracer. Proved deadly, the mere thought of it was enough to turn Sandy's guts to liquid. He was the only crewman with visibility straight down from his bomb-aimer's position. His dread was to see the silhouette of a plane before it fired its murderous round.

Upper gunner Frank brought with him a stigma of unpopularity. He certainly wasn't the most favoured of crewmembers. He showed a shy reluctance to pay his share in a shout and had been known to diddle the NAAFI ladies out of one penny for his cup of tea. He drank every chance he could but mostly at everyone else's expense.

And rear tail gunner, nicknamed Skinny on account of being tubby, was often referred to as tail-end Charlie. Skinny drank to escape the loneliness of being cut-off from the other six crew. Bill guessed his convictions might be flexible. Both gunners hated long trips as they'd little to do but watch and wait. Their turret spaces were narrow and confined.

'At least other crew members are able to move about a bit,' moaned Skinny.

'Skinny, you're such a poor bastard. You get to see nothing except where you've been but we'll make you useful yet.' laughed Bill.

Each new Lancaster pilot had to fly a single operation as co-pilot or flight engineer with an experienced crew. The problem was getting assigned, as no one wanted to carry a passenger. Bill went to Dispersal to have another closer look at a Lanc.

He stared at the retractable main landing gear and smaller fixed tail wheel and memories of Canada rushed back to greet him. Olly read his mind.

'Aye, try jumping out and turning the tail of that big motherfucker. Be like throwing fowls at corn.'

They laughed but the comment gave Bill a start. He pushed memories of Richard and *Alice in Wonderland* from his mind.

Tache tagged along. 'We suspect too many planes have been lost because they simply ran out of fuel. Others have exploded when they copped direct flak because they were carrying those blasted cookies.'

Everywhere about them was an awful distinctive whiff of fear topped by pitiless penalties of warfare. And then there was the waiting, waiting...

Tache became pessimistic. 'Base crews are rated as lasting about two weeks. That's about it, fellers, a lousy two weeks.'

Olly felt under siege. His stress levels rose. 'Aye, but could you naught soften it up a bit?'

'What, tell you stories? If that's what you want, you'll hear plenty of guff. Lancs have eight .303 calibre Browning machine guns. There's talk of heavier .50 calibre cannons to come. Conversely talk of cut backs to reduce calibre, and save on weight. Always talk, talk, and more bloody talk...'

Oh, fuck! Bill had hoped the war might not last this long!

'Most aircrew talk ranges from crappy aircraft to poorly trained ground staff. About the only thing happens quickly around here, is the removal of your kit when you don't make it back. Oh, and reinstatement of another crew assigned by OTU.' (Operational Training Unit).

Bill was dismayed. He'd realised he'd encounter problems. Never expected plain sailing, but all this negativity and official line guff demoralised him.

The Lanc looked huge by comparison to other flying machines. Specifically a bomb carrying aircraft it was intended to do nothing else. Stripped of everything heavy it was devoid of armour plate for protection unlike American Flying Fortresses, which lived up to their name.

'Only armour plate on a Lanc is a six inch square plate behind the pilot's head,' mused Bill.

'Aye, pilots can be shot to pieces. Hose what's left out, but naught will pass through the back of our fuckin' heads.'

Why this was, which military genius thought it up, Bill never found out. And on the massive wings were four of the biggest engines Bill had ever seen. Each geared to over 1600-horse power.

Bill let out a low whistle. He was impressed. 'A Spitfire's one pushes about 2000-horse power, but of course with remarkably less weight.'

'Aye, it's naught carrying upwards of twenty tons of bombs either.'

'You might be exaggerating, Olly.'

'My arse! Do your homework. An Avro Manchester was built to carry eight tons, and failed. The Avro Lancaster was built to carry a maximum of fourteen tons. Aye, word is we often carry in excess of that. A Tall Boy alone weighs twelve ton. And the new Grand Slam weighs twenty-two!'

'Fuck! I hope you're wrong. We'll never get off the ground.'

From his vantage point at ground level Bill shielded his eyes to better view his cockpit some twenty feet above him.

'It's a shame we couldn't wire a couple of those babies to a Tiger Moth,' he said to Mack as Olly went off to find his own plane.

'Better yet, if they slipped the stock-take,' said Mack.

The bomb bay was huge. Mack and Bill peered inside. An armourer was busy working on the photo flash camera to ensure the flare was timed to coincide with the impact of their bombs. It was designed to record the accuracy of the bomb-aimer's drop. It also verified their bombs had been dropped on target. Not elsewhere, and in particular not dropped somewhere innocuous like—in the sea.

'Gee, Skipper, we could graze a couple of sheep in here.'

'Bugger the sheep. It's ideal for our next social,' laughed Bill.

Later that afternoon Bill attended his first briefing with Tache's crew and experienced the chill of the Ops Room. It was nothing flash, a big draughty Nissen hut. Drab green painted walls added to the glum atmosphere on which sat two Air Ministry posters. They urged better teamwork:

One showed; *Don't be proud, stay with the crowd.*

The other; *In the likely event of baling out remember to tell them nothing but your name, rank and serial number.*

Some wag had sprawled a big *UN* in front of ...*likely.*

'All up it's a cross between a railway carriage and a giant ashtray,' observed Tache.

Intelligence gave advice. 'In the event of baling out over Germany, always try to board a train out of the area.'

Tache leaned in closer to Bill, as he didn't want to raise his voice. 'Forget their crap, Bill. That's bad news. Don't ever try to do that because German trains are heavily guarded.'

'Why do they say that when it's so obviously wrong?'

'Because British trains are unguarded and the fuckers haven't got a clue.'

As the nervous puffing of a hundred cigarettes mingled with the smoke of Bill's pipe, a thick blue haze formed in the air. More and more swirls of smoke joined gossip flying back and forth, but all eyes kept flicking nervously at the curtained end wall behind an empty dais. Swift bursts of nervous banter followed by short abrupt silences, then strident laughter at any joke that under normal circumstances wouldn't be considered at all amusing.

While they waited some scribbled letters home to parents, wives or girlfriends and all hoped those last thoughts so hastily penned to paper might not be found in their lockers to be posted next day.

When the CO entered the room and made his way to the dais, there was a mixed reaction. Most of the men stood-up and saluted. Many either didn't get up at all, or rose so slowly as to only appear to go through the motions.

Bit strange, thought Bill. Although he's a fuckwit, perhaps we should show more respect?

On cue an officer raised the curtain to reveal a large map of Europe. The target area was configured with ribbons. Those in crimson designated their route in. Green was the route out. Bill surmised their inference being dangerous to get there but safer to return home.

Handed a billiard cue the CO pointed his marker at the target and the banter died quickly.

'Oh, no! Not that fuckin' place again,' whined someone near the back of the hall.

'Gentlemen, there will be a quarter moon tonight with heavy cloud cover,' announced the CO.

A low moan and expletives emanated about the room but he continued, 'I expect a high degree of accuracy. Bombing height will be 22,000 feet.' He moved his cue to point at the last leg marked to the target. 'On your last heading onto the target you shall have the jet stream immediately behind you. This should give you a favourable tail wind. Your ground speed will be 290 mph.'

Someone blew a loud raspberry. The CO never faltered.

'Might that be right this time, Sir?' asked someone from the middle of the pack.

The question remained unanswered as another voice interjected from the back. 'It's about time we changed the route, isn't it, Sir. It's all too familiar to Jerry fighters.'

After a pause and momentary fiddling with pencils and talismans, section leaders took over for relays of times and bomb loads, which were expected to be regular.

Tache leaned in close to whisper in Bill's ear. 'Watch out for irregular bomb loads.'

'What's irregular mean?'

'So fuckin' big, you can't close the bomb doors properly.'

Wireless frequencies were handed out on flimsies followed by marshalling procedures, which included times for running up engines and take-off.

Most bomb drops were scheduled to take place in the middle of the night from about 2300 to 0100 hours. There was no exception. They were all to fly the same headings towards a target in Germany. Only height and departure times separated them. They hoped their target would be asleep and not expect them.

'Biggest danger,' said Tache, 'is flying into each other. Mid air collisions.'

Flak was predicted to be light to non-existent, and then on their return journey they were to jettison propaganda leaflets around Paris.

If the RAF's bombing techniques were in their infancy, so were their leaflet distribution skills. The message, brainchild of a bright spark in Whitehall, informed Germans how much better off they would be to sack Hitler, vote for Churchill and return their armed forces peacefully to Germany.

As if the German people had any more influence over the outcome of this extravaganza than their English counterparts had, thought Bill.

There were supposed to be no German night fighters, minimal flak, and everyone home safe and sound after about six-and-a-half hours flight duration.

The atmosphere was frigid. Bill felt he could have sliced it with a knife. He'd never seen a better example of command being drained away on what to him appeared a tide of despair. Loud talk mingled with expletives before they filed out to the Mess for their fresh egg pre-operational meal known as *the propaganda meal*. It comprised a fresh egg and anything else cooks could obtain from a farm.

'It's excellent food considering the meagre rations civilians get,' commented Bill.

'The only reason,' said Tache cynically, 'is if we do bale out and get taken prisoner we'll be able to tell the Krauts how well the RAF feed us.'

Before take-off the plan was changed. Bill was to go along for the ride and not in place of the flight engineer. He had nowhere to sit and was told he would have to stand, which would be tiring and uncomfortable.

Payback for the transfer application!

Regulation silk and woollen underwear were intended to insulate against extreme cold and help protect from fire. Then three pairs of socks, one silk pair first, followed by heavy wool, and fur lined flying boots.

Next were long johns, battle dress trousers, singlet, shirt, heavy woollen jumper, battledress jacket, scarf, and three types of gloves, silk, wool and leather. And in case anyone took a midnight swim a standard Mae West life jacket.

An air of improvisation hung over the hour. Most felt they were dressed like Michelin Man but it didn't keep out the cold. At about 20,000 feet the floor of the aircraft was so extreme, contact caused frostbite.

'Not good recommended practice to stretch out against the frozen floor,' said Tache sarcastically.

The crew had a last cigarette. Bill puffed anxiously at his pipe before he tapped the last of the unburned tobacco out on to the frozen ground. Tache had a nervous pee against the rear wheel for luck and they clambered aboard. Take-off was delayed.

The powers fired a red Verey flare to signal the op was scrubbed then changed their minds. Crews were supposed to wait in their planes for the start-up engines signal but then they heard delays might extend another hour.

'Fuck this!' said Tache. 'I'm not sitting here waiting for these dickheads. Anyone for a drink?'

They all filed out and walked through a convenient gap in the fence behind *Dispersal* into the back yard of the local pub.

After one round they were told, 'Please come back, chaps, it's on again,' so they clambered back on board. Bill had had a whisky so he felt a bit jolly.

'Full power to the engines,' Tache said, as they prepared for take-off.

Flight engineer leaned forward. His left gloved hand pressed the vibrating throttle levers hard against their stop. Tache held the Lanc hard against the brakes.

Their fuselage shook and reverberated from the power of four Merlin engines at full revs. Tache's eyes were locked in the direction of their flare path. A bright green light flicked on and quickly disappeared. That was his signal to go.

Tache released the brakes and held on tight to both sides of the control column. They gathered momentum as they bounded down the runway. Bill began to worry. He could see they were not getting as much power as he would have liked. Tache sat rigid and continued to fight the controls to correct any miniscule of variation left to right. He tried to keep them to where dead centre should be.

How hard should it be to paint a white or yellow line?

They thundered on, ever closer towards that great dark unknown at the end of the concrete runway. Any moment they might disappear into the black void beyond their lights. Bill knew take-off was the most dangerous part until they reached enemy territory. A lack of concentration by Tache, any slight misdirection, even a drop of one iota in power from any one of those four big Merlin engines could spell disaster. Their laden Lanc could veer off the runway at any point. If that happened nothing could stop them from ploughing through the unstable frozen mud or snow. Even if their undercarriage didn't collapse under weight it could result in an explosion by their thin-cased 4000-pound cookie bomb. Almost certainly the demise of another crew wrapped in sheets of flames.

For a moment it was as though their tyres were reluctant to un-stick and lift.

At times such as these, thought Bill, it's hard not to question FOB's gravity as a fickle bitch. But then with little runway to spare, Tache managed to get her off the ground.

Barely airborne and their undercarriage was retracted to avoid drag. We're on our way, thought Bill. And not a light is visible anywhere.

They continued to climb over blacked out England. When they crossed their coast at Beachy Head they still needed altitude.

The glimmer of water below was a clear indication they were headed in the right direction. According to their Navigator they should cross the French coast twenty miles south of Dieppe at about 8,000 feet.

Other than the glowing stubs of other Lancs' exhausts there was nothing to see.

With only the thunderous noise of their engines for company Bill peered ahead into the inky black void to try and estimate the distance of exhausts. He knew flames from their engines were anything from two to three feet long but what he saw was distant glows or pinpricks of light.

German fighter pilots wouldn't need to be overly skilled to aim their cannons somewhere between four such spectacular markers, he thought.

Their navigator kept breaking radio silence with his new compass bearings every six to ten minutes. He made it sound so easy. Their changes were standard dogleg procedures although hardly necessary as so far there was no sign of enemy fighters.

'Where the fuck are we?' asked Tache.

'Over water, Skipper,' came the swift reply.

Bill felt nauseas. He hoped it was airsickness due to him not flying the plane but as he became worse he thought it might have been something he'd eaten. Everyone else appeared to be all right. He removed his dentures and disappeared into the Elsan toilet, which at forty degrees below zero was not a pleasant place to be. A serious danger if you needed to sit down, naked buttocks in contact with bare metal. Bill spent his next few hours in and out of the Elsan, his head rested on his arm as he retched.

Weather was contrary to forecast. Somewhere near their target, darkness gave way to haphazard flak. Shells exploded high and low, but Bill saw little of it. All Bill saw was what was directly in front when he was crouched in the cockpit. To him it looked an impenetrable wall. That crews were expected to fly through it again and again seemed illogical.

All in all that flight lived up to expectations. It was almost a walk in the park.

There were few searchlights over the target and as the raid became scattered visibility was worse. Unbeknown to Bill their navigator felt fragile. A survivor from another crew in an earlier raid, he'd been showered in shards of white-hot shrapnel after a mortar shell had left a gaping hole in their fuselage not far from where he sat.

'That style of history tends to make a man edgy,' Tache said later.

If that wasn't enough to unsettle him, the next day he'd witnessed a badly damaged Lanc touch down on its home run. He was at Dispersal enjoying a cup of NAAFI tea when the Lanc hit the runway and did a belly flop. Then as its undercarriage gave way beneath it, it rolled to one side, broke up and burst into flames. Crew trapped inside were unable to escape as flames gained hold. Engulfed in a mass of dark red fire it looked a terrible way to die.

Back in the present he led Tache onto their target but he was not the only shaky one. For Bill who had returned from a frozen bowl of spew, thoughts of death might have come as welcome, if not too permanent, relief.

Everything seemed to be going well until a Lancaster ahead to port quarter was hit by flak. Illuminated in a sheet of orange flame before it became a ball of fire, it suddenly seemed to curl over to starboard and go down in a spin. Had it been higher and back a bit, there was a real risk it might have collided with them. Tache held the Lanc steady.

'Bomb doors open,' called the bomb-aimer.

Bill felt a great rush of air around him and knew everything not fixed down would be airborne about the fuselage. There was a wait of about a minute.

'Steady, steady—bombs gone!' They fell like crows swooping down on a flock of dead sheep, before Bill felt the plane lift without their additional weight.

'Bomb doors closed,' called the bomb-aimer.

Tache continued to hold their plane steady for the photoflash.

Then someone broke radio silence with, 'Let's go home via Paris, Tache.'

Navigator gave Tache his new heading for Paris and hoped his calculations were correct.

Somewhere in the vicinity of the Eiffel Tower they reached their drop zone.

The bomb-aimer began to cut strings that held bundles of leaflets. Each contained batches held together by rubber bands, which he then fed out through the flare chute.

Their Lancaster's powerful slipstream did the rest. Freed from their elastic band restraints individual leaflets were sent raining down in their thousands into empty streets and gutters below. That, or if the rubber bands didn't break, they'd act as a brick through someone's roof.

Pissed off but delighted to be alive Tache told his navigator, 'set a short cut without doglegs. Let's high-tail it for home.'

Tache realised his bum had gone numb and he wanted a break. 'If you don't fly this aircraft before we get back, they might make you come and do it again, Bill.'

Bill took over the controls of the aircraft and immediately felt better.

Tache gave him some advice. 'When you approach a target, Bill, it's my experience you can do little to avoid flak. Drive on through the blockade with everything crossed and hope for the best. I've learned though, the heavier the flak the less likely there'll be enemy fighters. No flak and watch out, the bastards will come at you from everywhere like bees from a beehive. Once you've dropped your bombs it's up to you. Get out of there as best you can. Thank Christ Bomber Command switched to night bombing. Less accurate but safer for us.'

'They care about us?' Bill sounded incredulous.

'Fuck, no! Couldn't give a toss but they lost too many planes.'

Bill flew the Lanc the rest of the way home and landed with only one unceremonious bounce.

'Not bad. Think you've got the feel of her now?'

'Thanks, Tache, I'll be right.'

At de-briefing tables some crews were visibly upset, others cried. Turned out they'd flown in at lower altitudes and been badly buffeted.

'Good job they weren't underneath me when we dropped our load,' muttered Tache. 'They'd have been fuckin' crying then.'

After de-briefing the crews met up in the Mess. Tache summed up.

'We were lucky tonight, Bill. Three didn't make it back. Doesn't always work out that way.'

Crowds that lingered in hallways seemed de-regimented. But for Bill the boundaries of the hours had blurred.

Next morning promised a day of crisp, cold air and blue sky.

Bill prepared to go up with his new crew to fly circuits and bumps. He left others to enter through the main crew door on the starboard side of the fuselage while he and Mack pulled themselves up into the body of the plane through the bomb bay. There on its roof Bill found Joe already in his seat under their expansive see through Perspex canopy. Bill squeezed past him to his pilot seat on the left. There he sat on a raised portion of floor and felt like *Alice in Wonderland* after she'd swallowed the cake marked *Eat Me* and begun to grow. As if to accentuate this Joe sat to Bill's right on a collapsible lower seat. Dials and switches were behind him and to his right. Where he sat he appeared as only two-thirds Bill's height.

Oil temperature, oil pressure, and fuel selectors to monitor complex fuel systems, and about twenty gauges had to be constantly monitored to ensure vital smooth running of their aircraft.

Bill let out a low whistle as he fondled the joystick. 'Not much joy in it,' he mumbled to Joe.

Joe laughed but didn't get Bill's point. 'It's never the same as in their text books, you know.'

Bill sat for a while and pondered. This was what they'd trained him to do, to fly this monster. Boot camp, Canada and even Sophie suddenly seemed far away and yet in terms of time alone none was too distant at all.

He ran his hands across the controls and as he did he closed his eyes as FOB had told him to in Canada. It was as if in Braille some hidden message might be relayed. He contemplated what lay in front of him. An uncertain future, he was sure. Did he even have a future?

As he sat and took stock he became aware of smells. There were mixed odours from fabrics of the plane around him. He smelled an over-ride of fuel, but there were others. Paint, and sweat perhaps? And dirt.

He looked at hastily applied coats of fresh paint and wondered what, if any, calamities might be covered over. He decided not to share his thoughts with Joe but he'd been right. It wasn't the same as in the books and their lectures. It was nothing like it. Somehow the drawings and diagrams oversimplified. Made it seem too clinical, too easy and straightforward.

After familiarising himself again with his new cockpit he went off to explore the rest of the plane.

The plane layout had Navigator Ben positioned facing to port with a large chart table in front of him. An instrument panel showed him their airspeed, altitude and other details required for navigation. All mounted on the side of the fuselage above the chart table. Bill noted he was further shrouded behind a blackout curtain where he could use a light to work. They had the GEE receiver as a navigation system.

'Aye, my cubicle makes me feel like I'm sat in a wee coffin, laddies.'

'Possibly we all are, Ben. But what are your thoughts about the GEE?'

'Aye, it's a bloody good system, Bill. But calculating our position between pulses can be a wee bit tricky.'

Bill wished he could go back and hug Captain Hawkins. What a treasure he'd turned out to be.

'If you need to brush up, Ben, let me know. I spent time with the GEE in Canada,' offered Bill.

Ben's face broke into a broad smile, 'Aye, thanks, Bill, I appreciate that.'

As the only non-novice aircrew he tried to look grateful. His job was not easy. A small anxious man of accuracy but rarely cheerful he was expected to guide Bill mile after mile through darkness of blacked out countries in wartime conditions. After a process of doglegs, he needed to put him over a point on a map where they were supposed to drop their bombs, and then guide him back again, using different doglegs to try and save their lives.

Radios for wireless operator Mack were mounted to the left-hand side of the chart table but faced the rear of the aircraft. There sat Mack, nicknamed unkindly by Ben as 'Seal Basher.' A name everyone else refused to use.

In theory Lancs had advanced communications systems but Mack preferred to use Morse code. He could listen rather than broadcast as enemy night fighters could home in on advanced systems.

'No surprises there,' muttered Bill.

Mack's Morse code rating was up to a respectable twenty-five words a minute and rather than concentrate on tasks at hand he seemed always to hum or sing the latest Sinatra number.

Their radar detector supposedly lit up whenever their plane was tracked by enemy radar but became disconcerting when triggered by false alerts.

'If in doubt turn the fucking thing off, Mack,' suggested Bill.

Mack nodded in agreement and exercised his sardonic humour, 'Don't worry, Skipper, I'll not be using it often.'

On his left hand-side was a window, used for visual signals for celestial navigation and above him their astrodome, used by Ben.

Down below was Sandy the bomb-aimer, who lay prone in the nose of the aircraft. He had access to controls for their vector bombsight and bomb release selectors. He used his view from the large transparent Perspex style nose to assist navigator Ben with his map reading. A mat on the floor was where bomb-aimer Sandy called home. A blond man, hence the nickname, he was a perfectionist. Thankfully his job matched his nature.

Behind them were main spars for their wings, which created major obstacles for any crewmembers moving along the fuselage even when they were stationary, let alone in the air.

At the area of the bomb bay the floor stepped down to the bottom of the fuselage.

It was from there Frank accessed his mid upper gunner position. From the roof of the bomb bay he had a small retractable ladder to help him pull himself up into his turret. He had the most uncomfortable ride of any of the crew. Seated on a stretch of canvas slung beneath the turret, once he was in, that's how he stayed. Unable to move, he could only fidget for up to eight hours at a time. Ben was unsympathetic.

'Aye, but at least Frank has a bird's eye view. I get to see naught.'

'Best view in the house,' confirmed Skinny.

He was right. His upper turret had a 360-degree panoramic view around and above, 'But I never bother to look frontwards,' he said. 'Bill and Joe are paid to do that.'

Instead Frank patrolled skies above and to the rear beyond each wingtip and down, but only as far as he could see.

'Anything below me is a total mystery. I don't even get to see bombs explode,' complained Frank who always seemed to have an unlit cigarette in his mouth.

Rear of Frank's turret was the side crew door but on the starboard side of the fuselage. This was the main entrance to the aircraft and their parachute exit. But to reach it they had to climb over the main wing spars. In flying gear with a parachute on this was not an easy task even if the plane was level. If being manoeuvred, or otherwise in trouble, it was virtually impossible. Bill wondered how many uninjured men might never have reached the exit of a doomed plane.

Perched at the extreme rear of the aircraft, over the spars for the tail plane, sat Skinny. He was a big man with an even bigger personality, and a round rollicking voice to match. His position was fully exposed and he entered through a hatch where he was then entombed. He could only see behind them. As they filed out, ground crew arrived to start their preparations.

Fate was indeed ready to unleash her fury on the bomber boys. Statistics being compiled would be sifted and regurgitated by Air Ministry, before Bomber Command rushed to hush them up.

'Not good policy,' ranted Tache. 'Top brass don't want us to know about their fuck ups.'

Correct, but would be handy to know, thought Bill.

January 1944 had resulted in the greatest bloodshed ever for Bomber Command.

'Aye, naught summat top brass is eager to shout about. Nor want to be read about in Daily Telegraph at their breakfast table,' said Olly.

Insanity is doing the same thing over, and over, again and expecting different results...(Albert Einstein)

Chapter 17

BILL'S WAR

Bill was called aside from his crew and issued a government revolver. It was a seven shot Ruger .22 calibre intended to fit snugly inside his flying boot. * 8th Footnote

'What's this for?' asked Bill as he balanced the small gun in his hand.

'In case you get shot down, Flight. Might be useful to defend yourself.'

'You serious, Corporal?'

The Issues Clerk took on a wounded expression.

'I'm very serious, Flight. Yes, it's a gun, and it can kill.'

'Probably could,' said Bill as he weighed it in his hand, 'if I manage to hit him through the eye, or slap bang in the centre of his heart. Provided he's not wearing a shirt. I wouldn't want to miss and wing him, would I?'

Issues Clerk shrugged. 'It's not intended you hold off the entire German army single handed with it, Flight.'

'No. Nor pepper Germany with it from the sky. It's about as useless as tits on a bull. Our flare pistol would be more lethal. If I remember, I might carry it.'

Another spoke up from his job stacking shelves. 'Think of it as having a condom, Flight,' he said with a smile.

'A condom?'

'Yes, better to have one and not need it, than need one and not have it,' he laughed.

'That's funny,' said Bill. 'I'll carry the gun.'

Planes fuelled and ready, Bill and his new crew were about to embark on their first op together. His head was full to bursting, a font of knowledge. But first there was a briefing to attend. Bill joined his crew and sat furthest from the front.

He noticed Olly and Tache. Olly looked pensive. He also sat near the back and they nodded acknowledgement to each other as they took their seats.

'You feeling ok, Olly?'

'Aye, but it's all about to happen. Just as they said, but by gum, we're a long way from Canada.'

Bill received few looks or acknowledgements from other crews. Not obvious at the time but he was soon to realise, while friendships formed within his own crew might be boundless, there seemed little point extending this to other aircraft.

Some crewmembers from previous ops were paranoid about where they sat. They had to be on the same chair, or at least in the same spot, as previously. Mostly others understood and made way for them. An anxious airman, who wanted to sit in a particular place, was not a man to be reasoned with.

Superstitious, thought Bill. Not a bad thing. Might be a heap of crap but why tempt providence?

A few laughed nervously. Others pretended not to hear them. Some appeared panic-stricken similar to Olly. Bill wondered what his own facial expression looked like.

When the officer entered, again only a few acknowledged him. Bill felt he had a fist in his guts. He smiled and hoped he looked more confident than he felt.

One wag near the front called out, 'Just another glorious battle day, Sir.'

The briefing was that, brief and to the point; a map uncovered on the wall showed a munitions target in Germany. Coloured tape showed spidery routes to and from their target. Crucial information as to its location, known searchlight and flak batteries along the way, was given in fast relays, to be noted or memorised as best they could. As Bill looked about him, for some reason Duncan's inane question came to mind.

'Please, Sir, should we be writing any of this down?'

When the barometric pressure was given a long low moan emanated from among more senior crews who started to heckle the floor. Bill looked about him in amazement.

Had he missed something?

Next subject was flack.

'Promised to be light,' said an officer. He looked an unsmiling streak of misery.

There were more heckles and rude comments. Someone at the back made a loud raspberry noise and a hardened young pilot to Bill's left, with a pink neck and protruding ears, commented shrilly, 'So, the whole of Germany will be expecting us tonight. Isn't that right—Sir?'

Even the way he said *Sir* had a barb to it.

The assembly booed, cheered and whistled en masse. But the question remained unanswered.

Next were flight heights, times and bomb loads. At the end of every sentence the long streak of misery threw back his head not unlike a typewriter, resetting its line.

But Bill saw through their polished display. In the officer's eyes Bill thought he saw a deep fear of the unknown. *Don't know why. He's not flying over Germany tonight, we are.*

'Take-off is set for 2000 hours,' announced the officer.

Adrenalin started to burn in Bill's blood stream. This was it. They were actually going to go and do it. Drop hundreds of tons of bombs on Nazi Germany. Chairs scraped. Conversations started as everybody stood up.

Before each crew left the room they had to sign their lives away. They received a locked dark green metal box. The contents were demonstrated for those newcomers who hadn't seen it before. It was a survival kit. In it were a number of items considered useful if stranded in a foreign country. The chocolate bars looked tasty; there was a compass and other basics included a silk map of Germany and a Swiss style army knife.

Bill mused about the knife and pondered how the Swiss Army ever became the authority on such a thing as a multi-function penknife? Having avoided wars for more than a century, Bill thought the Swiss Army and their knife should be in good company with the useless pistol nestled uncomfortably against his ankle. Anyway, to dice his apple he preferred his smaller pocket penknife. It was far less cumbersome.

The best of the box was the small King's ransom in foreign currencies. Each airman let out a low whistle of surprise. There was enough money to buy or bribe passage for seven men to anywhere. But there was a condition. Unless the crew were shot down, on their return the box had to be handed over intact, or it would be instant court martial, a firing squad—the works.

The Issues Clerk chose his words carefully. He never said, 'When you come back,' or, 'when you return.' It was a case of, 'A signature is required,' or 'will be required.'

'Not too many crews haven't posed the conundrum, Bill, over countless hours in the Mess, of how to break open that box and abscond with the money without being caught,' said Tache with a giggle.

If any had, they were unknown. Bill decided should he bale out the money and chocolate would be stuffed in his pockets. They could go fuck themselves with their Swiss Army knife, and the compass.

And that was that.

Out into the late afternoon twilight and dampness immersed in the colour of lead, and to prepare for going to war.

There was no hubbub, no excitement, and no crazy chatter. It was different to how Bill had expected it to be. Back in Canada when he'd taken off there was an excitement as adrenalin pumped. This was flat. No one spoke, and when they did it was matter of fact. It was as if everyone was aware the job had to be done but no one wanted to do it.

Now was the time to prepare. A vibrant sense of strain lapped over Bill.

His rear gunner Skinny was the heaviest man of the crew. Bill looked him over.

'It might be a tight squeeze, him in the rear gun-port. Upper gunner Frank would have been better-suited, size wise,' said Bill.

'Only if that canvas seat would hold Skinny,' smirked Joe.

Each crewmember worked to ensure their success. Ben their Navigator spread out his maps to examine the series of doglegs for their route. They were rarely instructed to fly straight in and out. Their way on to their target was by constant zigzagging. He also might need alternatives should aircraft start going down around him.

Mack needed to co-ordinate their radio and frequency charts. Some bands were notoriously unreliable. Better for picking up BBC than what they required of them.

Gunners ensured their Perspex turrets were clean enough to see through, tested their hydraulic motors powered by relays in one of the plane's four engines, which allowed them to rotate. Skinny worked through 180 degrees while Frank had a full 360. They checked ground crew had not forgotten to refurbish their ammunition belts.

After Bill had completed his dressing extravaganza he went in search of WAAFs who prepared their sandwiches and filled thermos flasks with hot sweet tea. They also packed and handed out parachutes, although some crew preferred to pack their own.

The standard joke by WAAFs was, 'Bring your parachute back if it doesn't work.'

When flight take-off became imminent, he and Joe, his flight engineer, checked their aircraft over.

'An unusual sky's happening. Much appears to be going on up there,' said Joe.

'Good,' mumbled Bill. 'Let's hope the weather turns foul and we don't have to go.'

He might get his wish. Darker wisps had formed where otherwise white clouds had feathered into grey. Joe looked drawn. He had a face like a brown paper bag, but with his moustache it made him look flat and frightened. A lock of hair hung across one of his eyes in a sweep to cover a bald place on his scalp.

As suspense rose almost to fever pitch there was a last minute interruption before departure. A perfect example of British bureaucratic stupidity required a dispersal form to be signed. Bill was to accept possession of the tonnage of bombs they were about to drop on Germany.

Bill glared at the corporal. 'Do you want a signature from the Germans for accepting delivery of them?'

In the darkness Joe and Bill exchanged glances.

As each aircraft began running up their engines the noise became super intense.

Then in turn each Lancaster moved forward, as if eager to take its turn in line. Then waiting, waiting for its green light before—it lumbered down the runway.

Protocols demanded their CO stand stiffly to attention and salute each departing crew. He was a martinet of immense natural presence in the failing light. Had he smiled, Bill thought he might have resembled Gary Cooper.

What their CO couldn't see, from within the gathering darkness from the cockpits, was the myriad of forks, birds and bum signs aimed at him? Each plane followed nose to tail to take-off.

'Full power to the engines, Joe,' Bill ordered.

Joe's hand rested on Bills to help him hold their four throttles forward to achieve maximum revs. The noise and vibration in their cockpit was horrendous. It was helped a little by wads of cotton wool soaked with olive oil Bill had rammed into his ears while he waited for his flask. Sole aim to reduce the constant drone of those engines and protect him from the extreme cold at high altitude.

His ear drums suitably padded, any chitchat while discouraged, was virtually impossible. Soon the cotton wool wads in Bill's ears would be no longer soft and fluffy. They'd feel more like munitions wads. But that would be the least of his problems. What he and his crew faced were a lonely and frightening five to six hours. And to make matters worse he already missed his pipe.

Back in the Mess he'd watched two airmen play dominoes. As a French import, prisoners brought dominoes into England during the Napoleonic wars, and the game had caught on. Last smoke he'd had was a few puffs while he watched them. It was then he'd realised board games were not for him.

Back in the present; Bill took one last deep breath. No turning back now! He felt as he had just before he stepped out of cover with his raised bat and lashed out at Alf's knee.

He cleared his thoughts to concentrate. With the extra bomb load they carried he would need every yard of runway to get off the ground.

'Maybe this world is another planet's Hell,' Aldous Huxley (1894 to 1963)

Chapter 18

FIGHTERS ALL

With four Merlin engines at full revs the big plane lifted slowly skyward through the muted, grey evening. Immediately their landing gear was raised to decrease drag and Bill was able to reduce the flaps.

'Bit close for comfort,' muttered Joe.

'Good job you got our wheels up. We only just cleared that church spire,' shouted Bill.

Over the intercom a crackle followed by a nervous laugh. 'Aye, but an inch is as good as a mile in our business, Skipper.'

Below was dark with the blackouts. There were few visible landmarks and what there were, were hard to make out. Bill kept his eyes glued between his artificial horizon, air speed and compass co-ordinates. It was now entirely up to Ben their navigator to plot their course by dead reckoning.

At their designated flying height, with some moon, Bill could see outlines of other Lancs. But it was their flaming exhausts that glowed like beacons.

He hoped the hardened young pilot to his left at their briefing, the man with the pink neck and protruding ears, was wrong with his prediction about the whole of Germany expecting them tonight. I hope the Germans might not see us as easily, as we are seen by each other, thought Bill.

Sandy, their bomb-aimer, took control near their target, as exact placement of bombs became his call. He guided Bill onto their target by dictating speed, height and direction. Emphasis was on flight deck to hold the big plane on course and steady.

But it was worse for Skinny crouched uncomfortably in sub-zero temperatures. Incarcerated into his rear gun turret there was little room for movement other than rotation from side to side. His parachute hung on a hook inside the fuselage near his turret doors. He had no spare space. Frank, his head and shoulders in his Perspex turret, stowed his parachute on a rack below him in the fuselage.

At 20,000 feet cold ate right through Skinny and Frank. With only the engine noise for company and darkness their security blanket, they longed for fog or dense cloud. They also hoped they wouldn't have to fire their dreadful guns.

Irrespective of where Skinny positioned his turret, the icy draft cut right through.

He and Frank's additional clothes and their electrically heated suits were intended to combat hyperthermia and frostbite at higher altitudes. Their ceiling height was about 24,000 feet but any higher and their plane's performance dropped off dramatically.

They each had a set of twin .303 calibre machine guns. Speed with which they fired meant each had less than fifteen seconds continuous firing time. As part of their training they'd been taught to lightly squeeze off a hundred rounds at a time in short second bursts. But when their guns were being discharged often something dropped off their plane because of the severe vibrations. That or a gun might jam if it wasn't already frozen solid.

What worried the gunners more than running out of ammunition were *tracer* bullets. Every third round lit up in the sky like an incandescent candle. Intended to assist gunners with their aim, it had a negative impact as it announced in triplicate where they were in the darkness and helped make their Lanc an easier target for German fighters. At times they became difficult not to miss.

No surprise that tracer was an unpopular feature of armament with most bomber crews. Not unusual for armourers to remove tracer although this was against RAF policy.

'Fuck policy!' Frank had said. 'I want tracer gone! I'm the gunner. It's my call. Not some chinless wonder sat at a desk. It's bad enough when you see tracer all about you. What with that and being buffeted by flak, and with those searchlights up your arse, remove the tracer or I won't fire my guns.'

Lancasters couldn't outrun anything. They didn't have sufficient speed or manoeuvrability. With red flashes from anti-aircraft guns below and being over 100 mph slower than the slowest German fighter, if picked on they were sitting ducks.

Germans held all the aces when it came to firepower. The Junkers Ju87 was a twin engine light bomber with eight machine guns and four cannon all wing mounted and set to converge at a point about 200 yards in front. A good place not to be if it fired, Bill's crew agreed.

Let alone the Messerschmitt or Fokker Wolf, which were dedicated fighter planes. If an enemy fighter targeted a Lancaster, it didn't have much in its favour. About its only option was what flying school described as evasive action, which translated was a downward spiral, un-affectionately referred to by flight crews, as the *corkscrew*.

Their only hope of survival was if the enemy fighter broke away and looked for an easier target. But there were no guarantees in war. If the Lanc was followed, it stood little chance of survival.

Odds of Skinny or Frank bringing a fighter plane down with their guns were beyond mid-range in terms of potential. Flying at night, without a bright moon, visibility might be less than one hundred feet. In the split second it would take they wouldn't even see a fighter, let alone hit it. And if they did their Lancaster would have to stay straight and level while they aimed. As if that was likely to happen.

Not that their bullets weren't capable of bringing another plane down. With guns aligned to converge at 250 yards, punching out combined rates a minute, they'd hurt all right. But to hit their target was problematical.

'Aye, at night hitting a fast moving shape from a lumbering platform is equivalent of a blind man swatting a determined fly with naught but a pencil sharpener,' whined Ben.

So many crews never completed their op. Many a Lanc crew had jettisoned their bombs harmlessly over the sea. Some deliberately flew around for the duration in ever decreasing circles, to return home without ever venturing over enemy soil, let alone their target. Others who had been grossly overloaded dropped their ordnance in desperation to gain height.

Hence the bomb cameras and the importance placed on photographs taken of where bombs had been dropped. Any crew caught dropping bombs not where they should be came under exceptional scrutiny and punishment.

Bill's first op was his most difficult mentally. Somewhere over Germany in total darkness he came to terms with his lot. He simply accepted he was about to die. No question about it. Anything else would be a bonus. By writing himself off this way he coped better with his op and traumas associated with it.

To a man the crew dreaded corkscrew manoeuvres. Not only did it mean they were in harm's way, making a desperate run for safety and closer than a hair's breadth to death's silent kiss, but it was a gut churning way to fly. Particularly when most crew couldn't see where they were going and had absolutely no idea what was happening around them.

Bill would learn control of a fully laden Lancaster throughout this style of gyrating evasive action was horrific. Not recommended safe flying practice, yet their alternative wasn't pretty.

He hoped if he had to use the corkscrew manoeuvre Lady Luck might at least smile on him until after his bomb run when the plane would be lighter and easier to handle. It was not an easy task in a gyrating dive to pull the monster back from virtual free fall. Joe had other concerns. He worried as corkscrewing and subsequent recovery meant over revving their engines, which used too much fuel. He worked on fuel consumption of one mile to the gallon. A round trip was approximately 1,200 miles.

Lulled into a false sense of security their first few hours had passed uneventfully. Then everything changed. As they neared their target Bill thought this op must be his last.

Suddenly their enveloped darkness changed. The sky in front lit up. For mile after mile it was as if they were in a light show. In the midst of the action were sheets of white light.

Red balls of fire exploded. This was soon followed by pinging and clatter sounds as bits of exploded shells hit and bounced off their fuselage. Bill felt himself shrink down in his seat as if trying to make himself smaller. *To be invisible would be good.* He flinched, and contracted involuntarily, as if to be further away from the horror that surrounded him. He realised although it was cold, he was sweating profusely.

They were surrounded by heavy flak. Explosions burst from the defensive guns below. It was a Guy Fawkes Night fire, smoke and tracer like fireworks filled their sky.

Oh fuck! 'I don't ever want to see another fireworks night again,' Bill said to himself as he fought his instincts to turn and flee.

Stories back at Base were rife of crews who had survived anti-aircraft fire but there was no information, only conjecture, about those who never made it home.

Bill couldn't hear further away explosions, and hoped they wouldn't get so close he might. In appearance they were like giant puffs of light. There were catastrophic great sheets of white billows of light as shells discharged from batteries below became more accurate in their grouping to try and bring them down. Bill had a sinking feeling in his guts when he realised to those brave souls beneath them, he, Bill, was their enemy.

His comrades might be white knights in their own eyes, but they were black knights to those below.

Sometimes it was as if the light faded before a plane was visibly hit. Then powerful ground searchlight beams roamed around. They flickered on Bill's Lanc for a moment, and then moved on. Perhaps he was unseen but certainly not coned. Then another plane ahead of him was caught in their beam. It didn't flicker nor did it move on as it waited for another to hone in and cone with it. It was like watching the wind. Bill could only see her, by what she moved.

Being hit might be similar to destroyers depth-charging the crap out of a submarine, thought Bill.

From momentary light flashes, before shadows took over, Bill could see shapes and outlines of other Lancasters, in front and up higher. He saw another plane to one side but had no idea what was behind. He guessed Skinny would have a grandstand view and might have some stories to tell, if they survived to relate their tales.

There was an explosion not as far away as he would have liked and he felt an awesome shudder almost like a pulling power with a sideways vibration through his controls.

Oh, fuck! She didn't like that, not one bit. Only difference between being in a submarine with depth charges dropped from above and him in a Lanc was he had windows.

'Thank Christ, I never joined the Navy,' Bill muttered quietly to himself. 'Imagine being incarcerated in a tin can. At least in a Lanc I might see it coming.' At times it was as though some giant hand reached out to try and grasp them in flight. Then as they approached their target, the intensity and frequency of the barrage increased as they all flew into the same cauldron of fire and flames.

Off to their portside it was a different sight. A Lanc was hit. Caught in the blinding glare of searchlight beams up came the flak. Sparks from debris pelted Bill's fuselage and reverberated off their wings. Then he saw flak strike the starboard wing of the other plane. It took off the tip and peeled back the covering metal skin with the ease of opening a tin of sardines.

Something ignited, perhaps a fuel tank, and a river of burning debris and sparks erupted and trailed behind it. Anti-aircraft fire seemed incredibly accurate.

An engine was on fire. Flames looked strange, almost false, as they licked past its fuselage. Then there was another flash, and another, much brighter this time. The Perspex dome in the nose of the injured plane shattered and sparks and debris splattered everywhere.

'Bloody fuck!' The other Lanc plunged downward. It was like some giant hand had pulled a master switch. A patronising view of the disabled plane meant Bill was left to wonder if he was hit, might he look similar to others. As the doomed Lanc disappeared from his view Bill broke out in another cold sweat. He was scared shitless. The other Lanc could easily have been him and his crew. That awful tinny taste was back in his mouth again. This time he fought back the bile as it rose in his throat big as a golf ball.

What if the next bright flash of light took him, his Lancaster and his crew? Bill moved uncomfortably in his seat. He knew he was about to panic. A downhill turn was shaping up nicely.

'It matters not whether you win or lose; what matters is whether I win or lose'...Darrin Weinberg

Chapter 19

BOMBS GONE!

It could all end here, my dreams and hopes could vanish on my first op.

And an end to his coveted nineteen shillings and sixpence a day.

If he flew at about 10,000 feet they wouldn't need oxygen, which removed one variable. But the 20 and 40 millimetre guns could reach him. At less than 4,000 feet he feared impact from his own ground explosions. At the height he was if his oxygen failed he would perish anyway. But the 80 millimetre guns could get him. Any higher and the 120 millimetre guns might fire at him. Any lower and planes could drop their bombs on him. And always fears of a fighter plane like a Messerschmitt. Only good news was he hoped German night fighters were unlikely to be active in this maelstrom for fear of their own ground artillery.

Bill tried hard to push these dismal thoughts from his mind by simply telling himself, 'it's not a good night to die.'

Around him it was as if the fight had morphed into some version of a macabre ballet.

Fuck! It's bullshit! We're losing planes and we're not even over the target yet!

Intercoms rarely worked properly and when they did they were barely audible above the din of their engines. Bill needed to concentrate to hear the words. He dragged his wads of cotton wool out of his ears and tried to keep a strict modus operandi over use of their intercom. The on/off switch was located on their oxygen masks.

'No idle chatter,' he ordered. 'Don't anyone break silence and talk unless you have to.' He knew he sounded nervous.

If someone left their mike on others might hear their heavy breathing.

'Someone turn that fucking mike off!' shouted Bill. He hoped the near panic in his voice was not as noticeable to others as it was to him.

Occasionally navigator Ben would break in. 'Bill, turn our wee plane to port five degrees, now!'

Bill gritted his teeth. All he needed was a comedian, but Ben was rarely cheerful and it did help ease their tension.

Silence was resumed for about ten minutes when Ben interrupted again. 'Bill, turn the wee plane to port two degrees, now!'

Bill recognised a light tap on his right shoulder and feverish hand signals from Joe indicated Sandy had already taken up his bomb-aimer position below Bill's feet. Bomb-aimer was the laziest of jobs until they neared their target. Otherwise besides dropping bundles of aluminium strips, called *window* out of his small chute in the nose, Sandy had little to do.

When scattered the cut strips were intended to confuse German radar. Each piece, although not large, showed up as a beep like a plane on German radar screens.

Faced with thousands of blips, in theory, Germans were unable to detect which were real aircraft. Supposedly this ploy worked—at least for a while.

Sandy was the only crewmember who could legitimately lie down on the job and even go to sleep if he wanted. Other than occasionally manning their front guns, which were rarely used he'd no other duties until they approached their target. Then, and only then, did Sandy earn his keep as he took charge. It all came down to him, their bomb-aimer. Laid out flat on his padded mat he was ready for their final run in.

'Bomb doors open,' called Sandy.

When their bomb doors opened they hardly needed to be told. They felt the sudden telltale draught. Mack and Ben made a grab at anything not secured. That or if you needed it, chase it around the aircraft.

From now on the only movements Bill made were those ordered from Sandy his bomb-aimer.

'Right, right...left a bit...left a bit...steady...steady.'

Bill held the column so tight he was almost afraid to breathe.

Forward through their windscreen Bill and Joe could see dozens of puffs of light as they burst ahead of them in the blackness. Occasionally, they saw an instant splash of crimson at the core of a blast.

Bill's pulse raced. It seemed an impenetrable blockade. He knew he'd no choice but to fly through it. Waves of panic swept over him. He imagined it to be not unlike a woman's contractions while giving birth.

Fuck me sideways! They're closer together.

In unison his breaths became short and fast. He sucked more deeply on his oxygen. His mind began to clear, his panic subside. He hoped he hadn't peed himself, but if he had, no one else need know.

It felt an eternity to Bill as he held the plane steady. Sandy took about two minutes to line up their target below with his little illuminated cross on his bombsight.

Bill felt like a slab of raw chicken before it shrivels in the heat of a fry pan. When would Sandy release those bloody bombs?

The seconds continued to tick into another minute but it seemed like hours. Tension on their flight deck was palpable. Bill dare not breathe heavily or move for fear of not holding their plane steady.

'Just drop the fuckin' wee things,' said someone. Bill had a good idea who.

Suddenly, Bill felt a surge upwards. Now lighter his Lanc lifted skywards as if it gasped with relief without its huge consignment of bombs.

Beneath him Sandy had pressed his bomb release button and Bill heard those magic words he craved to hear.

'Bombs gone, Bill!' Bombs capacity was supposed to be fourteen tons. In effect fifteen 1,000 pounders and incendiaries had fallen from their aircraft.

And then...

'Bomb doors closed.'

Quickly their intercoms were turned off before night fighters could hone in.

Bill held the Lanc steady. They had to wait another interminable sixty seconds for the photoflash. Without those pictures Bill and his crew might as well not have bothered to be there. It seemed an eternity. To take his mind off the wait he calculated the cost of the ordnance they'd just dropped. When he figured in excess of what a detached house with a garden in central London would sell for, he stopped.

Then as Bill looked down through his window he saw through the cataract effect of grime an impact of bombs. He would never know if they were his or if another Lanc had dropped them. Loads of death was exploding on the ground. It was alight with multiple explosions.

Lancs usually flew in attack waves about twenty minutes apart. In theory this was to avoid collisions but if the ground defences hadn't expected the first wave, Bill reasoned they sure as hell had time to prepare for what followed.

A heavy gloved hand lightly touched his shoulder again as confirmation of the all clear. Their job was done. Bombs had been dropped and photographed. On target or close enough, but certainly on enemy soil and not as some crews had done, out at sea. All he wanted was to try and return home alive and in one piece. He reasoned to achieve that, first he needed to get away from their target area as quickly as possible.

'Turn wee plane west,' said Ben.

Bill was confused. Which way was west? There was a pause while Bill collected his thoughts.

'Turn the wee plane to the left, Skipper.'

Bill's vision was restricted. It was night with some moon. He saw a Lanc in trouble in front of him and another to starboard and slightly higher. They were breaking up. It was a terrible sight to see. As he watched he realised his eyes were full of tears for the brave young souls who were dying on board and he wondered who they were. Might it be Olly or Tache? He imagined their terror. There was no sign of any crew jumping clear. Other Lancs otherwise invisible in the blackness suddenly glowed as out of control. In flames they spiralled towards the ground.

Bill broke into a sweat as he heard in his head the screams of wounded men, some hysterical, all wanting to get out. Their lives would have flashed before their eyes as their brains protected them from the all-incumbent horror. He knew some in their death throes would be calling for their mothers.

Far and wide the scene unsettled him. Up close flak petrified and there was a dark red flash, which at its heart was the kernel of the explosion. Bill couldn't feel it, nor hear it, but it was as if he could taste it. It sickened him. A close call again pelted his Lanc with fragments of shrapnel. Closer and it would tear holes in their wings or fuselage.

He wanted to throw-up, but this was not the time. It was as if the entire sky was alight. Searchlights beamed in around him. As they shone about he noticed there weren't nearly as many reflections or shadows of Lancs as before.

Many were gone and that didn't account for what had fallen behind him. He would never know what happened to them and if brave men had managed to bale out. For the moment he felt like a trapped moth caught in a spotlight.

He knew Frank and Skinny would have their eyes peeled. Watching, scanning the darkness with their turrets ever moving around and about. A night fighter would see them before they saw him because of the glow from their exhausts.

Shit! He was a sitting duck.

If flak didn't get him a fighter plane most likely would. All around were chaos, death and disaster. He fought panic again.

It was as if he was out on a limb trying to be noticed. Every second felt like an eternity.

Their orders had them fly into harm's way. To plot a straight and level course through dangerous barrages of flak long enough for their bomb-aimer to line up the target. And what if he hadn't? Bill would have had no choice but to fly around and approach their target again. Or fly against their mainstream with ever-present risk of collision.

No fucking way! Not a second time. Fuck their instructions. There will be no second coming.

Suddenly, streams of projectiles rushed at him. If they didn't get him, it might be a mid-air collision. Bill watched a Lanc to port partly obscured by cloud when another Lanc came upwards from below. He saw it happen. They hit hard! Two Lancs, possibly still with bomb loads, erupted. A massive flash of combined yellow tinged flames as a precursor to an enormous ball of black smoke as both were obliterated.

His training kicked in along with his instincts. Bill made a decision. With his bombs gone, this was not a good place to be.

He broke formation and dived.

Not the corkscrew but a clean dive. Direct and fast. But not fast enough. Joe anticipated Bill's move. They exchanged some quick sign language before Joe lowered their undercarriage to accelerate their descent. With their wheels down they literally dropped out of the sky at about fifty feet a second. Bill shivered and hoped there was not a Lanc directly under him.

Thankfully there wasn't. He wondered how many pilots in this same situation had shit in their pants. His speed approached 300 mph, which was far more than a Lanc was built to do.

He knew his crew needed him. It was up to him and him alone to get them home and in one piece. At this speed they could still get him, but he could be hit and in a dive. He felt safe and kept going. As far as his crew knew they were hit. Poor bloody Skinny he thought, might he panic, even bale out? All the while he dived he remembered Saskatoon when the engine had failed. The G-Force was similar.

At Base there was a story told in whispers about a crewman who, at the end of his tether, gave way to his tattered nerves and jumped out of his aircraft without a parachute.

Bill shuddered at the thought but felt sorry for the poor bastard. He sympathised, but if death is imminent, there seemed no point in hastening it. He knew he needed to pull up, but not yet.

Lower, lower. Fuck their rules!

He glanced sideways at Joe to his right and saw a nod of agreement. Then his eyes glued back onto the altimeter. He began to level out, the engine noises changed, became more regular. Only difference between Saskatoon and here was the size of plane.

They were less than 2000 feet now. Joe raised their undercarriage. He could clearly see flames of fires on the outskirts of their target. Even make out details of roofs and buildings alight.

But wait, something was wrong. Terribly wrong! His height! He was too low.

The instruments! They're out! They're wrong!

'*Fucking hell!*' Bill was shouting now. He felt sweat break out on his forehead.

No. It's all right. I'm flying level but I'm low, terribly low, and so much lower than intended, I'm almost hedgehopping. Up, up, up, more for Christ's sake! Bill knew he needed more height. But according to the altimeter he was flying at 2000 feet. Bullshit!

That was not his height. Not even close. He could make out shadows and people as they ran below, vehicles and terrible chaos but then as quickly, suddenly gone, behind him.

He leaned back slightly on the stick. Best forget his stupid altimeter. He continued to climb to what he felt was a more comfortable height. Now he knew what had been meant at the briefing. They'd been given barometric pressures to set their altimeters and the rude comments that had followed. That and the flak. Jesus! If that flak had been light, he would hate to see heavy.

He looked down at the ground and estimated his height at perhaps 500 feet tops.

That would do. It should certainly rule out a fighter attack from below. As long he didn't fly into the Eiffel Tower, who cared a damn?

He focused on his artificial horizon as another hand briefly touched his shoulder and passed him new compass co-ordinates to home. For the first time since he'd taken off he felt a bit safer. He was flying light and fast and as low as he dared. Not an easy target for a fighter. Too low for flak, but possibly there was danger of small arms fire.

They'd need to be lucky or damn good, thought Bill.

Bill's bum had spent more time in the air than on his seat during the past hour. He flew on. His next problem would be when he crossed the English Channel. If only his instrumentation was reliable, but a mistake at low altitude over water and he would be in the drink. Flotsam or jetsam, he didn't want to be either.

Sky and sea became a single world that ended in an envelope of darkness. He brought his Lanc up a few hundred feet. No point in crashing to avoid being shot down. The thought amused him. He laughed out loud, but no one heard him because his intercom was turned off.

'First sign of madness,' he said to himself. 'I'm laughing at dying. Stupid bastard!'

Noises of his engines were a welcome relief. If nothing else, that meant he was alive. Previous feelings of loneliness, cold, and his constant fear of death had temporarily left him. He relaxed a little although he found it difficult to loosen his grip on the controls.

Fear of death might be his constant companion but he'd found combinations of slow deep breathing exercises with repetition of his cockpit checks provided some degree of sanity.

He looked confident to his crew. If only they knew how close he'd come to losing it.

On their return flight Bill handed over controls to Joe, which in itself was a carefully rehearsed routine to avoid any misunderstandings. Dark, cold and poor intercom communication could play tricks on a man's mind.

'Do you have her?' shouted Bill.

'Yes. I have her.'

Bill prepared to sip his hot sweet tea. Adventure enough juggling a thermos while wearing heavy gloves but he needed to remove one to tackle his sandwich. It was certainly a welcome change from their tea at home. Rationed by the Ministry of Food they went without milk and sugar.

Their Lanc droned on. Each crewmember in his way, coming to terms with fighting an awful guts churning war against an enemy they didn't know or understand and couldn't see.

Bill shifted uncomfortably in his seat and wished he might remain at peace with his haemorrhoids. Then another problem emerged. While Joe was at the controls his airsickness had returned without warning. Tea and sandwiches away.

He was never so pleased to see the English Channel.

Now he knew he needed to concentrate. He estimated from appearance of white caps glinted on the surface below there was barely 500 feet between him and the sea.

He was perilously close to the water.

If a fighter plane wants us now, I'll make him work for it, and might even take him with me. But there were no fighter planes. Bill remained at barely safe flying height above open water until he climbed to cross his own coast. He reasoned sustained tension had reduced him to an artificial calm. But no point in being misidentified by some clown as German and shot down by his own side on his home run.

Sometimes through mist and rain even in daylight he'd had trouble finding the horizon. Once over their coastline and without Ben's guidance, even in daylight, close packed English villages looked the same. As they sped over roofs at less than 2,000 feet he imagined people in bed and their initial panic at the sounds of a bomber, then the peace they might feel knowing a friendly Lanc was thundering home after dropping its bombs on the Hun.

He thought of Fred turning to Lily in the darkness and saying, 'another of our lads returning home, Girl. I hope to Christ it's Bill.' His eyes misted before the thought amused him, as he knew he would be rattling their roof tiles.

After landing Bill walked around the Lanc to look for any telltale signs of damage.

He found plenty of what looked like bullet holes or possible shrapnel towards their port wing tip, not far from an engine. He let out a whistle and said to himself but loud enough for anyone close to hear. 'Thank Christ the Krauts didn't get the engine or fuel lines.'

The big issue now was de-briefing and nothing much would happen until after that. Everyone assembled back in the same room they'd left a lifetime ago. A few crews were deeply moved and one man was in tears. Earthbound intelligence officers ignored any signs of emotion. Instead they gathered to tally their count as other aircrews trudged back in. Heavy with flying gear, Bill flopped onto a stiff-backed uncomfortable chair set before an unsmiling officer, whose pencil was poised. He wondered at their philosophy. To deny warrior comforts by making him sit in cumbersome clothes, before being asked myriads of disjointed questions. This confirmed to Bill they were aristocratic wingless wonders, without empathy.

He knew he would find no noblesse oblige here.

While they answered questions crews smoked, chatted and poked fun at intelligence officers as they waited for their camera pictures to be developed.

'What was the flak like? Was it light or heavy? Did you see any enemy fighters?'

The temptation to say, I don't know! You chaps should come and take a closer look for yourselves, almost overpowered Bill, but he remained silent.

Time marched on. Bill noticed there was no sign of the young pilot with the pink neck and protruded ears. It looked as if he'd not made it home. Olly was there, thanks be to Christ, he'd made it back. So had Tache.

Most agreed flak had been heavy.

'Did anyone see any Lancasters go down?' asked an officer.

They had.

'Aye, I saw two go down. They just blew up. Exploded! I didn't see any crew bale out...' Olly's voice broke as the last of his words trailed.

Another pilot spoke. 'I saw one go down. Flack got his port wing. His two engines were set on fire. The wing collapsed and they spun down.'

'Did you get their numbers?' an officer asked.

There was a stunned silence. Was he mad or stupid? Had he any idea what he asked?

Bill was about to speak up when he caught Tache's eye. He remained quiet. It wasn't as though anything he might offer would be conclusive. Which plane, which crew? It was all a shambles. No one present was sure how many planes had been lost. Their de-briefing complement was depleted on numbers at the pre-mission briefing. Bill calculated from numbers of chairs occupied, about half had returned. It was hard to tell as the room had been rearranged. A few planes might be limping home, a few lucky stragglers to come. Olly stared at the floor his uniform and face rumpled with the night.

'Aye, true losses won't be known for hours yet. So far, out of fifteen crews who set out, naught but seven have made it back.' From where Bill stood, he saw Olly was trying to fight back tears.

Navigator Ben had helped get their wee plane there and back and now they were free to wash-up, change clothes and adjourn to the warmth and comfort of their Mess.

Bill's crew were over the moon, but not without some criticism.

'Why had he dived? Had they been hit?'

'Yes, we were hit,' said Bill. 'Port outer,' but he was reluctant to say too much. He knew when he broke flight formation he'd gone against instructions. He'd been given firm altitudes at which to fly in and out. But right or wrong, they were home and in one piece. And their bombs had hit their target area, or close enough, which was a bonus. Bomb-aimer Sandy could take an additional bow.

'Your blood's worth bottling, Bill,' said Mack. 'Thanks for getting us home mate. Perhaps you're Arsey as well as Ballsy Bill.'

It was the stuff of legends. There were murmurs of approval from everyone except Skinny who had been entirely un-nerved by their dive and had considered baling out. But they were agreed; only thing they liked about flying ops, besides the obvious of returning alive and in one piece afterwards was their welcome home dinner.

Bill set about his favourite death row meal; a massive rare steak pink in the middle with real fried eggs, chips and trimmings.

'It's so raw, cook showed it to the grill instead of putting it on it,' quipped Skinny.

'He's a machine,' said Joe. 'For Christ's sake keep clear from his face while he's binging, scoffing and troughing.'

Tache wandered over. 'Pleased to see you made it back, Bill.'

'You too, Tache. How did you go?'

'Different! Our Navigator had a rough trip. His nerves were shot and we got lost.'

'Where did you end up?'

'Fucked if I know. Not sure what part of Germany. We got separated from the main bunch and couldn't find the target.'

'What was it like?'

'What? Staring into the darkness of a blacked-out continent?'

'Aye, we all do naught but that,' offered Olly.

Tache stared at Olly. 'I took her down low to try and identify where we were. Anything might help. Terrain. Buildings. All I could see was water glinting below me. It was a big river but on our map where we were supposed to be there wasn't any water.'

Bill who had finished his main meal prepared to eat an apple with the aid of his penknife and as he did his mind raced back to Canada. There his attempt to follow a river had almost cost him his life.

'What happened to the wee river?' asked Ben.

Tache glared at Ben. 'Suddenly we went from dark to light. Everything seemed to open up on us. We had searchlights, flak, and tracer coming at us from all directions.'

Olly laughed nervously. 'Are you bloody serious?' His voice carried, and Bill winced out of embarrassment for his friend.

Tache continued. His voice as cold as charity. 'Even saw one of their new blue searchlights, tracked with radar. If that latches onto you, you're in real trouble. We got out of there quick smart. I told our bomb-aimer, drop your fuckin' bombs quick smart, then we found our way home.'

'Aye, so you know naught for sure where you dropped your bombs.'

Tache looked at Bill. 'He's quick, your mate, isn't he? I've got no idea, Geordie. Perhaps some chinless wonder of an intelligence office might know from the pictures. One thing's for sure.'

'What's that?'

'It must have been enemy territory because the fuckers were shooting at us.'

'How's your wee navigator?' asked Ben.

'He's pretty shook up. He collapsed. Poor bastard is in hospital. Might end up in the loony bin. From what I can make out, the river we followed must have been the Rhine. Looks like we might have dropped our bombs on Dusseldorf.'

Ben let out a low whistle. 'You were lucky you made it back, Tache. Dusseldorf is heavily defended.'

Tache's smile stuck to his teeth. 'Tell me about it,' he said. 'Anyway I need another navigator to square the circle, preferably someone who can read a map. Anyone hears of anybody, let me know.'

They bought another round and thanked their lucky stars to be home and safe, but for how long?

Next day they were stood down because of inclement weather.

Again the meteorologists were incorrect but that was not unusual. They were wrong most of the time.

Bill thought, I wouldn't mind a well-paid job like that. It's the kind of work where if you're correct some of the time you get promoted.

Because foods were rationed outside Base, steak and eggs were unavailable except on the black market. This meant Bill's favourite feast was a treat of gargantuan proportions for any aircrew fortunate to return.

It remained a complete mystery to Bill, how Air Force cooks could turn yellow egg powder into a muddy brown sludge.

'Aye, what ends up on our plates is intended to resemble scrambled eggs—not sticky grey clay,' laughed Olly.

'Perhaps it's payback from all the young hopefuls who've been reassigned to the kitchens instead of flying their dream,' offered Joe.

Ben's face turned to the floor, his spine ramrod straight. 'Aye, some wee dream this has turned out to be. It's naught but safe anywhere but in the kitchens.'

Bill changed the subject and they agreed the cooks do a fine job.

'Aye, but best of all are those traditional home coming dinners,' said Olly.

'Yes, any time, day or night, but how do they know in advance, how many crews they'll be catering for?'

'Don't see how they can know summat like that. We don't know ourselves.'

They fell into an awkward silence while Bill tackled his steak. It was always a big slab of what looked like beef although on occasions he suspected it wasn't.

'Aye, might be horsemeat, Bill. There's summat of a steady flow of naught but neighing animals through those adjoining paddocks,' laughed Olly.

'Irrespective of origin, it's good to eat.'

They felt privileged eating fresh meat and it being cooked to their taste was a bonus.

'I don't give a tinker's cuss whether it's been mooing, neighing or barking as long as there's plenty of it.'

'Aye, tenderloin of rat, or summat by any other name, would go down all right, too, by gum.'

There were often more than enough, even for second helpings, if cooks misjudged how many might make it back.

'Provided we survive, it's a win, win situation all round.'

'Aye, you're right. There's plenty of everything. And to top it off eggs are from a chicken's arse and naught the stodgy grey sludge served up at breakfast.'

Principal down side food wise was an over abundance of boiled carrots, served with every meal. It was argued by the powers carrots helped aircrew to see in the dark. This was nonsense, but as they ate bucket loads of them they became a vegetable Bill disliked intensely. He swore never, ever to trust carrots.

Hot sweet tea slurped in vast quantities, was according to rumour, laced with bromide, as His Majesty recognised the side effect in reducing sexual urges of his fighting men. A claim vigorously disputed by the powers but not for one moment doubted by Bill.

'Their tea tastes like shit and its good effects are next to useless,' claimed Bill. 'Paint me cynical, but how come when I'm away from Base, even for forty-eight hours I get a horn in the mornings Rufus would envy and a sex drive to match?'

'Aye, Bill. I feel summat like a gorilla in mating season. But at Base my old fella hangs like naught but a limp lettuce.'

Size matters not...(Yoda)

CHAPTER 20

NINETEEN SHILLINGS AND SIX PENCE A DAY

A wind had blown up from the North Sea. It rattled windows inland and brought a cold wind from Siberia. Tomorrow, it might rain, thought Bill. Another day gone and he'd survived.

At least until midnight tomorrow!

Bill had become well suited to pilot's wages and took easily to winning additional respect from those around him. He realised to continue he had to stay alive. Survive. Meantime he decided to share his good fortune at home. He insisted Lily and Fred leave their rooms in Acklam Road, Notting Hill where the grim evidence of poverty was everywhere.

Bill hated the narrow crowded street. Both sides were five-storey terraced houses. In their day they were grand but now, to Bill, they appeared to lean heavily against each other, as if for support. Their height precluded any rays of sun, except briefly at mid-day.

'It's time to move. We need a better place out of this shit-hole,' announced Bill.

With his help they moved into new digs at 2/8, Barlby Road, Kensington. Their flat was upstairs in a row of two storey-terraced houses. With its own indoor toilet and kitchen it was a more modern, worthwhile experience that suited Bill well.

He laughed. 'I'm surprised Barlby Road has survived the onslaught of the Luftwaffe.'

'More than can be said for East End and Dockland areas, Son.'

Not far away they'd been almost entirely destroyed. Whether this was pure luck or attributable to the genius and foresight of the German war machine remained a mystery.

With an entire upstairs flat to themselves there was no comparison between their new world and their old, where they'd shared an outside toilet and a kitchen without running water. There were other mind-boggling improvements too. The tenement building in Acklam Road had had gaslight and no electricity; their only tap on a half-landing was a long way for Lily to carry potable water.

Each house in Barlby Road had its own small front garden and some even had a pivot hedge. But there was no basement. No iron railings separated buildings from the street. Bill liked that. He had an aversion to iron railings.

Upstairs they had their own toilet on a half landing, two bedrooms, large front room and a scullery kitchen at the rear, which overlooked a small walled garden with joint rights shared with the downstairs tenant.

Things were changing for Bill. At last he felt he was making them happen.

'We might keep a few chickens in the back yard, Girl,' suggested Fred.

'That'd be nice. We could have eggs or fresh meat once in a while.'

Lily loved flowers and in particular the scent of hyacinths. Not only their sweet fragrance, to her their presence meant re-birth. At Acklam Road there had been no garden, only a smoky backyard where sunlight never reached.

Fred had organised a small window box to stand on their sill where Lily could water and tend her flowers. She had a garden now, to transplant her beloved hyacinths into, if she wanted.

Their small kitchen was more a scullery. Somewhere to sit, relax and eat. Easier to keep warm it was dominated by their hob, a closed fire range. A deep porcelain glazed sink replaced their old shallow one shaped from a slab of stone. It now sat within a cupboard well adjacent to the hob with its own cold water tap.

Their large blackened kettle was filled with water from the tap, and lifted onto their hob with ease. As their fire was kept alight, the kettle sat and simmered. Lily was pleased.

'We'll keep the kettle on the go. There'll never be an excuse not to have a cuppa.'

Their adjacent draining board served as a stand for washing dishes, a rubbing board for washing clothes and a place for accessories for ablutions. To one side of hob sat Fred's tattered armchair positioned so his feet almost stretched to the firebox.

Everything about their new address from the friendlier Victorian wallpaper to their rear outlook was better. At their old address they backed onto noisy sets of railway lines. Trains rumbled past frequently. Steam locomotives hauled freight trains and passenger trains were electric. Occasionally an Express thundered past. When he was younger Bill remembered the difficulty of sleeping through the noise but had always enjoyed watching steam trains pass by. He came to realise the experience of his first adrenalin rush.

As the train approached noise gathered in momentum. It rose in a crashing crescendo like some great Ravel's Bolero. When it drew near, there was a billow of grey smut; smoke and steam blocked light out. The noise was tremendous as it hissed and spat. His chest tightened. It became difficult to breathe as the monster approached.

The colossus of iron and steel thundered past and even the solid concrete and metal footbridge further along the street vibrated. Gradually the noise decreased as the great Goliath of the track raced away.

Now their view was serene. No noise, no smut, no steam. Instead, an uninterrupted view of their neighbouring bowls club green. They could sit at their scullery window with their blackout curtains hooked back during the day, and watch genteel local members play lawn bowls while they had a cup of tea.

One shared bath for occupants of both flats was situated downstairs. Rented by a retired couple they saw possession of their bath as equivalent of ownership rights. An appointment had to be made about a week in advance to use it.

This option tended to be overlooked by Lily and Fred. They felt it was an embarrassment to impose on their neighbours, which often over-rode their need. Their scullery served not only as their kitchen and sitting room but also their bathroom. Hot water was mixed with cold in the sink or a bowl on the table in which they strip bathed.

Lily was delighted. 'I'll have tea ready when you get back, Bill,' she beamed.

'Ta, Mum.' But deep down Bill was upset. Lily had said, '...when you get back, Bill.' No one ever said that at Base.

Barlby Road was on a busy double-decker bus thoroughfare. Buses stopped close to a number of local places of interest; Wormwood Scrubs Prison, White City Stadium, Ladbroke Grove Railway Station, and the Gas Works, which they knew was there when the wind blew from the west. Designated a through road, occasionally an armoured military vehicle passed by. When it did, it sounded like someone in a full suit of armour was being dragged up the street screaming and kicking.

Lily turned her nose up. 'Our old flat in Acklam Road wasn't as noisy at the front. And when your dad did an eastern run as guard I sometimes waved to him from my window.'

This ungracious attitude of Lily's annoyed Bill but he bit his tongue. In an effort to ensure his parents moved, Bill paid their rent up front but attached a condition to his generosity.

'The second bedroom's mine and mine alone when I'm home on leave. And everything remains undisturbed, even my old cricket bat propped up in the corner.'

Lily was delighted. 'We'll call it Bill's room.'

Bill thought for a moment before his superstitions kicked in. 'No, Mum. We'll call it the little room.'

He pondered where he might put Olly's gift, the thermometer.

'What's that, Son?' asked Fred.

'A small gift from Olly.' Bill handed it over for approval.

'That's nice, Son. It's a thermometer. We haven't got one of them, and it's pretty too. Why not let me put it up on the scullery wall for your mum?'

'Sure thing, Dad.'

Bill's room became known as the 'little room' and his thermometer had pride of place beside their mantelpiece in the scullery.

About to see it was Lily's sister, Marge. She had a daughter, Susan, about Bill's age. Lily and Fred's main source of pleasure before the war had been filling out their football coupons on a Wednesday night. Their lynch pin on these occasions had been Marge. Together they tried to predict six correct numbers for Saturday's draws, after which they would become instant millionaires. That never happened.

'The great Lord of winning football pools moves in mysterious ways,' laughed Bill. 'A foolish way to spend money and a waste of effort unless you get pleasure from the game.'

Bill had never played Pools or any other numbers games and didn't intend to start. He was not a gambler in the true sense although daily he risked his life, which he reckoned was gamble enough.

Marge was a kindly lady of later years with a shrill voice a few octaves too high for Bill. Also she had an enormous mole in front of her chin and in his younger years Bill had stared at her mole, often with his fingers in his ears.

Some things Bill was told as a child remained with him for years. One was frogs gave people warts and therefore anyone with warts was likely to be a frog fondler.

Confused, Bill thought it was a wart on Marge's face and he labelled her frog handler extraordinaire. He wondered if Marge would mind if he took to her dark growth with the kitchen scissors but no one else, only Bill, seemed concerned about her mole.

Marge was a First World War widow. Her drab looking spinster daughter, Susan, had eyes for Bill. She'd always liked him even when they were children. But now Bill was grown up and a pilot, she could hardly contain herself around him. A small problem, Bill lacked interest in her.

'Possibly he has an aversion to becoming romantically inclined with his first cousin as Queen Victoria did,' said Fred.

Bill lacked respect for these people as they dreamed about what they'd do with the money if they won. Lily nursed her hot cuppa in her hand and stared at the faded wallpaper.

'I'd buy a nice little place by the seaside.'

Visions of armoured trucks bulging with money, queued up outside their little flat sustained her.

Whenever Bill was home on leave Marge and Susan's visits were more frequent. Susan intensified her longing for Bill, which at times became more than a little obvious.

'Oh, Bill, you are so strong and brave,' Susan said. Her eyelids appeared to be on a constant flutter. Lily was in favour of the match but Fred took Bill's side.

'You'd need to be wealthy as royalty to get involved with anything that barking mad, Son.'

Bill preferred to be outside when home. Often he wanted fresh air. In their back garden he'd play and bond with neighbouring puppy, Scruffy. The friendly little dog warmed to Bill in an instant. He soon became lost in its licks, tail wags and liquid brown eyes.

It was a cheerful sunny day for a change. Fred captured the moment. He took a picture with his Box Brownie camera

Sometime after Bill put Scruffy down, there was a massive bang. Fred screamed and took cover. Bill threw himself to the ground. It was a critical moment. Jerry had dropped an incendiary bomb. It fell to the rear of their garden right on top of Scruffy and with barely a yelp the dog was gone.

Showers of metal fragments ricocheted off the ground before that horrible 'Wooooof' sound of an eruption of flames.

Trained to react in an emergency Bill and Fred shouted, 'Don't panic, Lily,' and promptly ran into each other. With no way of knowing for how long it might burn once ignited, Fred raced inside to get a bucket of water.

Bill grabbed a spade and began to throw soil but in his ignorance Fred threw water over the bomb, which made it worse. Bill narrowly escaped being badly burned and turned on Fred. He had his hands clasped around Fred's throat when Lily stepped in to separate her two men. Bill swore abuse at his dad at the top of his voice. 'Jerry has failed to kill me. Why are you, you silly old bastard, trying so fucking hard?'

Whatever was simmering below the surface had risen up in Bill, yet again.

Fred did his dying swan act and went off to make a cup of tea, but this exposed only the tip of the iceberg of Bill's bad feelings towards Fred. The water over the incendiary fire and subsequent loss of dog Scruffy were merely the catalysts.

'Their fight was probably about a lot of different things at once,' said Lily later. 'Bill liked Scruffy more than he cared to admit. After that he never owned a dog.'

Bill was angry his dad had never strived to achieve. That Fred was a guard on the railways further irritated Bill. He usually kept their relationship to one of manageable dislike.

'You're not even a bloody train driver, Dad. What's the matter with you? Have you no aspirations at all?' raved Bill. His dad's lack of ambition was unacceptable to Bill.

It is easier for a camel to pass through the eye of a needle, than for a rich man to enter the Kingdom of God...(Mark 10:25)

Chapter 21

REWARD TIME

After a while Bill began to resent their slap up celebratory meal after an op. He began to see the feast as the RAF's pompous gesture for them dropping bombs on Hitler and returning planes intact.

'It's a means-justifies-the-end crusade, Olly. Each CO country wide is content.'

'Aye, you're right, Bill. Naught but to go and do it all again, by gum.'

But first there was leave to take. That was the system. Bill had not yet realised he could never beat it. He would never win. Individuals were too small.

'Their chain of command is always right, even when it's wrong.' moaned Bill.

He had a few days leave owing to him so he used his free travel pass home to Lily and Fred's new flat in Barlby Road. He'd settled well enough into their little room, which had become his bedroom and he hungered for the home it represented more than ever.

During the day he did nothing much. He sat outside until the cold drove him into the scullery, yarned and smoked his pipe. Sometimes he thought of Eric and when he did, he felt sad. But Bill philosophised how he was not the same as he was before his ops. He might have been handling the spectre of death better than Eric had, but then he reasoned by comparison Eric had been a mere child.

On the wireless Vera Lyn belted out We'll Meet Again, and the White Cliffs of Dover. It was as if whenever the radio was turned on, she was always there and Bill grew sick of her. He preferred wartime hits like, Bless 'em All, and Kiss Me Good-night Sergeant-Major also played many times a day on the wireless, but more cheerful.

Lily and Fred couldn't get enough of Vera Lyn's songs and that irritated Bill.

Their wireless was enormous size wise to the point of cumbersome. It was an ugly wooden box with only two knobs; volume and tuning and if you twiddled the tuner far enough you could beam into almost anything, even German stations. Designed and built according to manufacturer Pye, to stand up to wartime conditions.

A bold claim indeed, but probably true, thought Bill. He looked it over. He'd known the wireless since he was a child and they were correct. It would have survived anything but a direct hit.

The day had become pleasant with more sunshine edging through dense clouds than normal. Bill looked out from their rear sash window and analysed the cloud formations; cumulus down low, strata higher. As he lost himself in his thoughts the sun warmed the closeness about the room and Bill's gaze settled on the garden.

There was an orderliness about the scene he realised he'd missed. The flowerbeds had been weeded and watered and a battered old watering can stood beside the garden tap. Each new row of seeds had a little bamboo stick with a push-in label written in indelible ink. They were stuck in everywhere in clumps, which indicated where something worthwhile had been planted.

He noticed Lily's hyacinths were back in their original pots barricaded behind bird wire to keep them protected from their chickens.

Lily came in to the room. She looked tired.

'It's hard to sleep with the worry and the air-raids, Son.'

Bill nodded in agreement although he was not in sync. He started to notice things that might have been there before, but were beyond his reach of reason. And his dad, Fred, looked more drawn than usual. He listened to what they said to each other. Chitchat without purpose. Somehow it was as if their conversations were banal, and there was something else, predictability taken for granted by Lily and Fred, something no longer shared by him.

'Your aunt, Marge, will be round again Tuesday. She likes our new flat, and we'll have another nice cuppa. You should be here, Son,' said Lily, 'Marge hasn't seen much of you since you made pilot.'

Bill felt an anger rise inside him.

Had his mum gone mad? Had she lost all sense of reason and gone silly in the head? Here he was, flying ops over Germany dropping bombs. Did she think it was a walk in the park? Bill fought back the urge to blurt out an insult, to tell her. He realised his mum either had no idea, or was trying to act normal. But what was normal? Was it bringing a rational moment into their irrational and crazy lives?

Would Bill still be around on Tuesday? Who knew? Least of all him so should he play along? Bill felt he could no longer take being alive for granted. His parents were at risk. They could be bombed out at any time but Bill felt the likelihood of him not returning from his next op, was far greater than Jerry finding Barlby Road in the dark.

'That's nice, Mum. I'll try to be here.'

Lily beamed, and then she looked embarrassed, turned away and wiped her eyes with the corner of her apron. In that moment Bill realised his mum did know. They all knew. They realised the reality about his poor life expectancy, but like mums the world over, Lily had trouble coping with the madness of it all.

Anything that resembled calm was held onto, like a drowning man in his death throes as he reaches out for a trailing rope. Bill felt a lump rise in his throat. He turned away so no one would see his tears.

Fred sensed his son's predicament. 'The garden's looking good for this time of year, Son. Stan and I have planted some tomatoes and those chickens are coming along a fair treat.'

Bill looked hard at his dad's face. A moment before he would have torn him apart for uttering such a trite comment. Now, he could see the value in it.

'That sounds good, Dad. Why don't we go and have a look?' Father and son left the room together, and Lily had her first opportunity for the day's proper good cry.

Out in the garden the moist smell of wet foliage after the night's rain had a pungent odour.

'Where do you suppose we'll find those chickens, Dad?'

Fred stared at his son. 'You'll find them somewhere around your feet, I'd expect, Son.'

Bill looked down. True enough one that resembled a storm trooper was pecking at his shoelace.

'We could do that one for a start,' suggested Fred. He aimed a sideways kick. It squawked and flapped about a foot in the air, then took another nip at Bill's boot on the way down.

'We'll definitely do him, Dad.'

'You're on, Son. I don't like its attitude either.'

Despite a hearty meal of chicken Bill struggled to fall asleep. He found relaxation difficult. His mind needed to unfold the day's events and when he drifted off his dreams became nightmares. Coned in searchlights from below there was no escape. His Lanc was rocked by flack and began to jump sideways from explosions. He was hit! His oxygen was cut off and he lost command. His Lanc rolled over and went down into a spin. They were in trouble. In chaos he fought to regain control if only for long enough to allow his crew to bale out. He saw a starboard engine burst into flames. Small arms fire pelted the fuselage. Joe was gone, a gaping hole where he'd once sat. Windscreen and controls were splattered with blood as flames erupted and spread throughout the plane. It was terrible. Ben and Mack screamed when they became trapped by flames.

Richard, so calm, appeared from no-where. 'Here, Bill let me help you put on your parachute to bale out,' he said. Then he looked cross. 'I told you it might be important one day to jump, Bill.'

Anne was there in a state of panic and shouted. 'No, Richard, don't let me fall. Shag me one more time before we die, please.'

That switched to him watching another Lancaster go down when flashes from a Messerschmitt's cannons took out their port wing. Horrific screams of crew as they perished, he saw Skinny firing from the rear gun turret when they blew up. As he fell down into the abyss, Skinny reached out both arms to Bill and sobbed, 'Bill, I was scared and you never told me to bale out.'

Life is not measured by the number of breaths we take but by the moments that take our breath away...(Anon.)

Chapter 22

MADNESS APRIL 1944

Oh, fuck! Bill swung his legs out of bed to get a glass of water. 'I can't take much more of this crap,' he said softly to himself. Not only was he fighting the RAF's titanic battle but reliving each and every twist and turn while asleep.

He found Fred sitting in their darkened scullery. His only light the flickering fire-box at the hob. Reflections for some reason brought back memories of the basement at the Church Hall where he'd first met Sophie.

'Didn't know you were up, Dad. What's up? Mum all right?'

'Yes. Thanks, Son. I couldn't sleep. I kept tossing and turning. I thought I'd sit a while.'

Father and son sat in silence. Bill wondered if his dad had ever experienced sex as he had in Canada with Sophie. Perhaps he had with Lily? The thought amused him. Fred's words rang in his ears, 'If one woman isn't enough, then 10,000 women won't be enough.' Growing up Bill had wondered a fair bit about that...

Somehow the echo of Fred's comment stayed in his brain but felt hollow, as now he had no woman of his own. He thought of Richard, and wondered what he might have said, 'Terrific old boy! 10,000 women you say. Bring them on, chaps. Don't hurry. We'll take our time and do them all!'

Bill wondered if Richard was dead. Then again, even if he wasn't, he might as well be. Fred stood up and stoked the hob. He filled the kettle and placed it back on the hob, and when Lily surfaced a few hours later, there they sat. Bill too afraid to sleep, for fear of revisiting his awful nightmares. And Fred sick with worry, whether his son might see another leave.

Sipping their hot strong tea, Fred's from his saucer, they both had one thing in common. They wished this terrible war would end. Bill's experiences had grown progressively worse. Recent raids on Berlin had claimed more than seventy Lancs. Jerry had wiped out entire RAF squadrons. Another raid at the end of March had totalled a further ninety-five, who failed to return. In one night alone, more aircrew had died over Germany than in the whole of the Battle of Britain.

Olly's comment came to mind. 'What if they know naught about this nor running any war, but it's our lives on the line?'

After breakfast Bill took a stroll to the shops. There wasn't much happening. Most were empty. A man selling winkles and cockles from a barrow-stall stood on the corner.

A policeman stood beside the stall and took a winkle. He worked the meat out of its shell with a pin from his lapel, and then tossed the empty shell back onto the heap when the stallholder wasn't looking. Bill almost said something but decided not to.

What few vehicles were on the roads were fitted with headlamp covers with crude slits to allow a glimmer of light. Enough for minimal visibility for night driving the powers said, but they looked comical during daylight hours.

Because of the shortages of petrol, there was little traffic on the roads otherwise carnage from accidents at night would have been catastrophic. Road direction signs and many street names had been removed. Another brilliant idea conceived by the powers intended to confuse possible invaders. 'If there were no signs anywhere, the enemy would become disorientated and confused,' they said.

When Bill returned home he mentioned this to Fred who scratched his head in bemusement. 'Being lost in our strange land, Son is an interesting concept. Does that mean they wouldn't be able to make sense of a map, or forgotten how to read a compass, perhaps?'

Lily sniffed. 'Not able to find Buckingham Palace in time for tea with the Queen more like.'

'You mean if they can't find 10 Downing Street or Chequers at Chartwell they'll feel impelled to call it off and return home,' Bill said with a smile.

'That's about the measure of it, Son.'

'Withdraw and go back home,' sniffed Lily. 'That's what the Germans will do.'

'We could be over thinking it, Mum.'

But such a brilliant idea had locals even more confused than usual. Before this happened they'd some idea where they were, even how to get elsewhere. Now they had none. Many of them were found wandering around in a bewildered state:

'Excuse me, mate, is this White City Road?'

'Well, yes mate, I think it is. Or, at least it was yesterday, I think, or perhaps they've changed it.'

'Well, mate, do you remember what it was called last week?'

'Wouldn't want to put you wrong, mate. Better knock on a door and ask someone who lives here!'

And then:

'Oh, no dear, we're not sure are we, Stan? You see this is my sister's place. We were bombed out last week. We knew the address there. That was Cavendish Street before signs came down. Does that help you at all, love?'

Countless hours were lost in utter confusion.

'Perhaps,' laughed Bill, 'If the Germans do land they might be able to help us. Tell us where we are or how to get somewhere else.'

Hand delivered telegrams with important messages, were late or never arrived. The fourth series of Ordnance Survey ¼ inch maps began to be issued in 1933 and proved successful with motorists. They were comprised of about twenty overlapping sheets, which covered all of England, Wales and Scotland at a scale of four miles to the inch. One of its values was each road being identified by its Ministry of Transport number. It was an innovation that stemmed from 1791 when the military required maps of southern England, should France invade England.

'Not much has changed over the years, has it?' sighed Fred.

'Now instead of Napoleon we fear Hitler,' sniffed Lily.

English people could be seen out and about at road intersections, their maps spread on car bonnets, their heads gyrating. Between studying maps every which way, they searched the heavens for any glimmer of sun to try and discover which direction was east.

Bill laughed. 'They look like schools of demented pilot fish. Then they turn and return their map in a desperate attempt to find any direction at all.'

'Like which road might go somewhere useful,' sniffed Lily.

'If someone happens along they're gold,' said Fred.

'Doesn't matter if they have no more idea than anyone else,' continued Lily. 'Anything's better than nothing.'

'God only knows how many tried to find Chartwell and ended up in Hastings,' added Bill.

'Even military intelligence is in trouble.'

'I'm not surprised, Dad. A misnomer if ever I've heard one. As for trouble, it's a long queue. How can the military be intelligent? And I should know. I work for them.'

'Bamboozles the army on their manoeuvres, that's what.'

'They're incapable of tracking a wounded elephant in the snow. That's the funny part but the real worry is they have unlimited budgets and powers to boss us around.'

'The solution is to enquire of the nearest publican at any good tavern and rest a while, as they should know something meaningful, like where they are,' laughed Fred.

'Never mind finding their way,' said Lily, 'what about those awful sirens?'

The air-raid sirens sounded off at the most inconvenient times, usually after dark. The official word was to take cover or else. They signalled death was on its way.

Lily and Fred had a choice, but how much stemmed from Fred's involvement with the Home Guard, Bill wasn't sure.

'So, what are these great mates of yours in Dad's Army offering you for further home protection?' Bill asked sarcastically.

Fred shook his head. 'The choices are a bit limited, Son. Two types of shelter. Anderson and Morrison.'

'So, outside in the cold wet garden or crouched under a kitchen table,' observed Bill. 'What have you decided?'

'We've both coming, Son. Stan downstairs wants an Anderson because with his bad back he can't climb stairs and he hasn't room for a Morrison on account of their kitchen being too small,' said Fred.

Bill out stretched his arms in hopelessness and spoke to his mum.

'So, wet and miserable, outside in the Anderson. If you use it don't forget to take a bucket with you, to bail out water.'

Lily tightened. 'Yes, we know, and candles and matches for light but not a glimmer must show.' She produced a torch. Old with weak batteries, it projected a beam that even in total darkness wouldn't make the slightest difference. 'As long as no-one sees the light.'

'Not much chance, Mum, at least not with that.'

'Stan said he's going to get his son around to line the inside of the shelter with old newspapers.'

'You're putting a Morrison shelter in here as a kitchen table, are you?'

'More comfortable inside than outside under bent pieces of corrugated iron, Son.'

'But you won't fit one in here, will you? They're big, you know. Hard to get into a small room and then again not as safe if you get a direct hit,' queried Bill.

Bill had his doubts about Anderson shelters. Rarely did they work out well.

Most were installed incorrectly, and then left vacant only to be used for storage.

'They're not much chop, Dad.'

Pushing back in his chair he impersonated Stan's missus calling out from downstairs. 'How long have you been out there then, Stan?' he mimicked.

'Oh,' Fred caught on to Bill's little game. 'I'd say about six hours, fifteen minutes, and thirty-five seconds last time I looked at my watch. But I'm not really counting, Faye!'

They laughed and Lily refilled the kettle.

'Not many problems around here can't be sorted out over a nice cuppa.'

'Downstairs inside the cupboard under the stairs is another option,' suggested Fred.

'Supposed to be the strongest part of the house. But to fit your mum, Stan, Faye and me downstairs in one small cupboard would be a squeeze.'

'Don't anybody fart you mean,' said Bill.

'Shush,' said Lily, 'that's rude.'

'It's rude but true. What you want to do is push Stan and Faye out into the uncomfortable Anderson shelter, and you two go hide in the broom cupboard.'

'There or the tube station, Son,' said Fred.

Lily was horrified. 'What, down into bowels of the earth. I don't want to go down there. Not onto those London Transport train platforms.'

'Why not, Mum? At least you'd be safe down there. No bomb of Hitler's would get you down there. Not now the powers have stopped locking the gates to keep people out.'

'Long walk, Son,' said Fred. 'Closest is Ladbroke Grove, a fair old gallop from here. Why did they do that?'

'Some theory the powers had, that if people could hide down there they may never want to come back out. That worried them. The war effort might be slowed up.'

'That and no guarantee any buses will run.'

'It's not bad down there, Mum. Not once the trains stop,' Bill reminded her.

'You been down there, Son?'

'Only to catch a train, not to sleep.'

'Well, I'm not going down there,' said Lily. 'Picking my way between bodies snoring and passing wind.'

'It's like being at the bottom of a whore's handbag down there in them tube stations,' said Fred.

Lily squirmed. 'Watch your language, Fred, there's no call.'

'Just like that pretty little girl we knew from Woollies. What was her name?'

'You mean Sylvia. She was a dear little thing.' Lily settled down to tell her story:

'She used to go down at Queensway Tube Station during air raids. Chose there because they were deepest underground, she said. Picked a spot each night, usually between people who looked familiar. She'd settle down snuggled up nice and comfy in her eiderdown until trains stopped running. Then after the trains stopped she noticed this strange shuddering.'

Bill looked vacant.

Fred explained. 'What your Mum means, it took Sylvia a while to work out. Then she realised.'

'Realised what, Dad?'

'You know.'

'No. I don't know, Dad. Not until you tell me.'

Fred looked flustered. 'She realised blokes were doing things.'

Bill looked vacant.

'Men were doing what comes natural to them,' said Lily. 'Now do you understand what we mean?'

'Oh, under the blanket activity, you mean.'

'You could call it that,' said Fred with a smirk. 'All around the poor girl men were enjoying their own company. Now I've said, it. That's an end to it. Do you understand now?'

'No wonder she was ill at ease. Poor girl, it must have been embarrassing for her.'

He cast his mind back to Sophie and Anne. There was no-way he could even begin to broach the subject of them with Lily or Fred. But the thought made him smile. He wondered what Anne might have said or done if young men had been masturbating under their blankets around her.

'Mary told me at the grocers. That's why she stopped going down there. All those faces she recognised,' said Lily.

Fred was supportive. 'She knew if she saw them in the street she wouldn't be able to look them in the eye,' he said.

'So instead of going down to tube station she sits under the kitchen table in her little flat.' Lily's voice started to break, as she came close to tears.

Fred rose to comfort Lily. He put an arm around her and held her close.

'Don't cry, Girl. There's nothing we could have done.'

'No, but those Germans,' sobbed Lily. 'They didn't have to kill that pretty little girl, did they? What earthly advantage was her death to them?'

'Your mum's right, Son. She copped a direct hit. Right on top of the house, even took the neighbours out. Poor cow would never have known what hit her.'

'Another unfortunate casualty of war was what newspapers called it,' sniffed Lily.

'But it's so wrong, Fred. All those poor innocent young people.'

'In England there's never been mercy for the poor,' grumbled Fred. 'As far back as Tyburn Tree, they've paid for everything, even their own broken necks.'

They agreed.

'But it's happening almost every night, Mum.'

Fred was dismal. 'Yes,' he said. 'Any night during the Blitz there can be several warnings before the all clear sounds in early hours. Then survivors emerge from their hiding places. You see them rub sleep from their bloodshot eyes, most in disbelief. To see if they have a roof over their heads or not.'

Lily noisily exhaled. 'Or pick through rubble for days to find something of value.'

'Nothing to eat, nothing to wear, nothing at all and peak hour times ten,' summed up Fred.

They were right London barely functioned.

'A fortnight I can only look back on through deeply etched snapshots,' sighed Bill.

But at least they were all still alive.

Masses of people attended memorial services, flowers and wreaths when available were dotted all over London, and everywhere crowds gathered. For Londoners this was the era of the bulldozer and the backhoe. But it would have taken every machine in hell to dig one enormous pit big enough to cover the annihilation in the East End.

We have seen the enemy and it is within us...(Pogo)

Chapter 23

FARCE

Small focus was given to the 'Home Guard' by His Majesty, which comprised men too old to join up and fight overseas.

Bill had a question, 'Why is that, Dad?'

The older man was reluctant to speak about it.

'But it doesn't make sense, Dad. If they're prepared to send kids as young as fourteen to die at the front line, why get a conscience about killing off men, old enough to drag their scrotums?'

'It's nothing to do with conscience or right or wrong, Son. Comes back to things we've seen. That's why.'

'I don't understand. What's that got to do with it?'

'Well,' said Fred, as he rolled his smoke. 'We older blokes have seen a bit of battle. We know what it's like to face a bayonet charge or have to attack a machine gun nest. Kids haven't been through that, they're still gung-ho, whereas an older man might hesitate, hang back a bit. Not want to rush in.'

'The bastards! You're telling me these Generals strategise to send in kids because they're too green to understand what dangers they face.'

That's about the size of it, Son. Put another way, they don't want me there telling them what shrapnel does to a man when he collects it full force in his guts.'

Bill let out a low whistle. It started to make sense. The powers had said men were too old to be pilots at twenty-eight. Perhaps that was to do with maturity and wanting to stay alive?

'Not that the Military strategy is right, Son.'

'You mean they should conscript older men?'

'Of course they should. Think about it. We're older, we're bad tempered and irritable. My back hurts and I'm always up early to pee. Young blokes don't like to rise too early. I figure I might as well kill a German arsehole, while I'm up. That might even make me feel better.'

'Perhaps you have a point,' laughed Bill.

'If captured I wouldn't spill the beans, because I forgot where I put them. Name, rank and serial number could be a real brainteaser.'

Although Fred was too old to be classified as cannon fodder in this catastrophe, he and Lily had developed a life-long dislike for people of German origin. Being decent caring people and Fred intent on doing his bit, he'd joined as a volunteer warden in the Home Guard.

'Your dad paces the streets alone at night,' said Lily proudly.

Fred's main task was to check the 'black-out' curtains had been properly drawn and if not, report any glimmer of light to those inside. Particularly dangerous when an air-raid warning had sounded and he was too far from shelter. Many a miserable cold wet night found Fred huddled somewhere; behind a brick wall, or cowered in some entranceway, as bombs fell lighting up the night sky around him. He was too scared to make his dash for safety.

'If there aren't enough terrible things happening, Dad, I've to worry about you keeping the home fires burning,' said Bill. He felt a sudden sorrow for the older man.

'I've seen some bloody things, and I'll see a few more yet.'

'You're not alone, Dad. They're always warning of invasion. If they're really serious don't you think they should give you a gun?'

'His Majesty refuses to issue Home Guard with guns.'

'I don't know why, Dad. It's not as though guns are in short supply. It's ammunition we're short of. And you might never fire off a round.'

Fred looked smug. 'Stan downstairs, he's got a gun. I've seen it.'

'What sort of gun?'

'It's an old service revolver from last war. He took it from a dead officer.'

'If the old fool fires it, it'll probably blow up in his hand,' laughed Bill. 'Why don't they give you Sten guns?'

'Oh, no, we wouldn't want anything fancy.'

Bill snorted. 'Fancy! I don't know about that but they're safe enough. They're jammed most of the time.'

'His Majesty's unwilling to issue your dad with anything much at all,' said Lily. 'He even supplies his own hard hat.'

One thing Fred and his colleagues were issued from His Majesty's diminishing stockpile was a pretend gun. Fred laid it proudly on their kitchen table.

'I don't like guns in the house,' announced Lily.

'You call this a gun in the house, Mum? It's hardly a gun.' Bill picked it up and gazed in amazement. 'It's a piece of timber.'

'Never mind, Son. It looks like a gun.'

It was a large piece of wood, and if Bill blinked hard then kept his eyes closed, it resembled the crude shape of a Lee Enfield rifle.

'Imagine confronting the enemy with this, Dad,' laughed Bill. 'Eight foot German death machines land by parachute, armed to the teeth with the best weaponry known to man-kind. They fire from their hip as they land and you're there with this piece of wood.'

'I could fire planks,' Fred said, straight faced.

They laughed.

'And Stan's service revolver,' added Lily. 'Don't forget he's got a real gun.'

Bill looked at his parents in amazement. 'I suppose you'll shout at him, will you?'

'That's right, Son. Surrender you bastard, or bang, bang—you're dead!' shouted Fred.

'Dad, the more I see and hear, more I wonder, really I do. What will you do when he doesn't stop? Call out, oh, fuck, he's not stopping…!'

Lily chose to ignore Bills potty mouth. She laughed along.

Bill had always suspected his dad and fellow wardens might not be the sharpest tools in the shed but they were doing their patriotic bit for His Majesty, no matter how unbelievably contrived. This angered him.

'We get government training, Son,' said Fred smugly.

'What? Along with their issue of wooden bang-bang sticks?' The more he heard, the angrier Bill became. He knew there were the usual fiddles. Some local dignitaries prospered from this delusional drivel, but that his Dad was actually being taught to shout, 'Surrender, you bastard—or bang, bang, you're dead…!' And in such a cogent and threatening manner as to convince the monolingual foe into thoughts of his self-preservation and disarmament, left Bill cold.

'Have you gone stark staring raving mad, Dad? I'm sorry, Mum. But this is preposterous. What you're telling me is that,' Bill held up the piece of misshapen wood, 'this is the only officially inspired defence of our once great nation.'

'You're missing the point. The way they explained it, it made perfect sense. Don't forget it's night time and dark. Hard to tell whether the gun's real or pretend.'

'Ignoring how dangerous this would be, Dad. You intend to out bluff a German paratrooper. Remember he's extremely well armed, well trained and disciplined. It's lunacy.'

'It might work, Son,' said Lily. 'In poor light or with cover of night.'

'You, Mum, are as crazy as Dad.'

When Bill calmed down he saw the funny side. 'No doubt the Hun trooper would appreciate the warning. Realising the cavalry was on its way, he'd refrain from following any of his Teutonic training, and in the face of such total devastation, promptly surrender.'

But humour aside, Bill wanted to know, 'Where is the British Government's 'duty of care' to protect you poor bastards? Obviously all lives can be gambled cheaply, but their own. What patronising piffle is this?'

'Taken to its logical conclusion of reality, sorry I mean—stupidity. Had you pretended to fix bayonets, your cry might be; 'surrender, you bastard, or stab, stab—you're dead…!' Bill held his head in his hands.

Fuck me. This is Dad's Army.

Fred warmed to the humour. He stoked the hob and put more water into the kettle. 'If we pretend to throw a hand grenade, how's about; surrender you bastard…or throw, throw— kaboom, kaboom, you're dead…!'

Lily laughed until her sides ached. 'You're under my feet you two. Go on get out of my kitchen. Go and do something useful. Tank, tank—move, move…!'

Tis all a chequer-board of nights and days, where destiny with men for pieces plays... (Omar Khayyam)

CHAPTER 24

TOP SECRET

At Base everything was supposed to be top secret. What Bill and his crew couldn't work out was why, when they went into their local village pub for a drink the entire bar knew they were on an op before they did.

'The pub always gets it right, ten out of ten,' whined Skinny.

'This is unacceptable,' stormed Joe.

'I agree,' said Bill. 'The need for secrecy has been rammed into us from day one. We're constantly reminded how loose talk costs lives.'

'Especially as it's our lives at stake,' added Frank. 'When I arrived, I was told my life expectancy was four weeks. Some greeting that was.'

'Better than Tache,' quipped Bill. 'They told him two weeks!'

'Aye, it's as if informers are everywhere. Nobody and naught can be trusted,' said Ben.

'But how do locals know more about what's happening than we do?' asked Sandy.

'I feel like my testicles are being used as footballs,' hissed Bill.

Prior to being briefed on their next op they left Base for a stroll and ended up at the local for a drink. There they were greeted with, 'you blokes hitting Harburg tonight. That right, Bill?'

Bill's crew were devastated by how well informed they appeared to be at the local pub. They even knew it was Harburg and not Hamburg. As a result they started a witch-hunt of mammoth proportions at Base.

'Someone's letting the side down but who and how to nab him? Joe raved.

Bill stormed in to see the CO. 'Do you know yet if we're flying tonight?'

The CO's face flushed red.

'That's classified information.'

'Not in the fucking pub it isn't.'

'Watch your mouth, Flight.'

Bill stared him down, his eyes spitting venom.

The upshot was top brass talked of leave cancellation to stem the breach.

'They've got the wrong end of the stick. It's not us doing the talking,' raged Bill.

Later, after the dust settled the publican put Bill and his crew out of their torment when he whispered the solution across the bar.

'Ah've summat to tell thee,' he said. And the baffling mystery of how the pub fraternity and every local knew what they weren't supposed to know was solved:

'Aye, at end of runway is an old church.'

They knew that. It frightened the life out of them every time they took off. To raise their undercarriage to miss was a point of extreme urgency.

'Bloody typical of the RAF,' ranted Bill. 'They've extended an old airstrip to take the heaviest bombers imaginable. Fair enough, I suppose. But here we are with bomb loads so massive our bomb doors have to be locked open on take-off and what have we got?'

'A church. Slap bang right at the end of the fuckin' runway,' grumbled Joe.

'Fuck me! It's the tallest building for miles around!' added Bill.

'You know, it wouldn't have taken too much imagination to get around that problem,' said Sandy. They all agreed, even the publican.

'Aye, a further example of so-called RAF's brains being naught but tarnished,' enforced Olly.

'So, it's agreed,' laughed Skinny. 'If anyone over shoots on take-off, they'll most likely find themselves in church.'

Everyone laughed except Joe. 'That would be funnier, Skinny, if it wasn't so bloody true.'

'But what has any of this got to do with secrecy?' asked Bill.

'Aye, but this old church, along with many other places of worship,' continued the publican, 'has a fine tall steeple with an adjoining spire. And at the highest point is affixed a modern red warning light, my young warriors.'

'Hmm. Probably to remind low flying aircraft it's there and provide an additional perch place for tired fuckin' birds,' quipped Mack. Again they agreed with much nodding of heads.

Granted the connotation of a red light, often associated with houses of ill repute and now, perched on high and on a church, was humour not wasted on lads at the local.

'Light is always switched on prior to announcement of any op. So everyone in the village knows your squadron will be flying later that night. Admittedly, we're not sure where,' admitted the publican.

'Well, not yet, but give them time and they will,' added Mack.

'They'll know that little gem before we do, for sure,' said Joe.

'You'd be surprised what gets talked about in this bar.' The publican leaned in closer. 'And I'm sure no-one means any harm. Your ground crews when they refuel have paperwork as to distances and how far you'll be flying. When your armourers check your bomb racks and machine guns they have a pretty good idea what target, according to paperwork. Even WAAF's who prepare your flight rations.'

Bill's crew let out a collective whistle of surprise.

But the red light of conjecture continued to be switched on for the duration of Bill's ops from that particular airfield, even after the 'powers' were made aware of the ramifications. Bill came to the conclusion, as did his crew, little they did or said was ever likely to influence those in charge of making decisions.

Joe laughed. 'In true RAF style, there's little else to do but mix a stiff gin and tonic,' he said.

Bill settled into a level of acceptance about the stupidity of his employer the RAF but started to become ultra superstitious. He tried not to be, but couldn't help himself.

At briefings Bill found his brain turn to soup. He decided he was past it. Sometimes only the scrape of chairs or louder buzz of conversation snapped him from his trance. He found himself always wanting to sit in the same chair as last time because he'd survived. At times his superstitions were silly. Over the top even to him; same number of teaspoons of sugar; three in his tea, but the same sized spoon rounded full to the same degree. At one stage he even carried his own spoon to ensure no variation. He rose to wash and dress at the same time each morning and completed his ablutions in the same order. Bill had become as paranoid as those who had been before him. His only saving grace, he wasn't alone with his irrational ideas. Whenever anyone spilt salt in the Mess there was always a rush to throw a pinch over each person's shoulder, lest they tempt the Gods.

Before an op there was reluctance among all crews to talk about their impending mission. Some sat around and wrote letters home, which they placed in their locker until after their return. Others drank excessively but there was always an air of apprehension.

Bill had drunk to excess as he'd found great solace in a bottle of Scotch. Now he tried to curtail it. Initially it had buoyed him up but then as with most alcohol it had the opposite effect. It depressed him. The emotional slump outweighed his feeling of well being for a while but he grew to dislike the after effects. The wretchedness of the hangover, the vomiting, or worse, his inability to throw up.

When he slept at Base it was as though he was in a room half the size of a broom cupboard. The walls seemed to close in around him. Then only another drink could quieten his demons.

'Some men can drink too much and sleep it off. They become generations of hardened piss-heads. Others throw-up and can keep drinking, but for Bill it ended with dragging his sorry arse around for days and with a king sized hangover no one wanted.

Not much fun feeling wretched so he tried to moderate his drinking and like Greek warriors over 2,000 years before him, they cut their wine with water and spoke of many things except that which was foremost on their minds lest it tempt the gods. History showed he was not alone in his superstitions.

Bill usually enjoyed loud cacophonous singing around a piano, any piano, and it was even better with an enthusiastic concertina player in tow. But there was a false gusto before an op. A rousing chorus of Blaze Away lacked that special ingredient when sung hours before going to war.

Same song later after a safe return brought tears to the eyes of everyone who sang it. Drinks flowed. Once Bill became so enthusiastic he offered to play the piano.

Next morning he held his head. 'God knows how much or what I drank. But whatever it was, it was fantastic. I even played the piano afterwards, I think.'

And so Bill opted to perfect his cow-like ability to sleep while standing up at briefings, particularly if he'd missed his opportunity for a favourite chair. He reasoned it was better to stand in the correct place than sit in the wrong position. Halfway through began to feel like the millionth he'd attended and he found it harder than ever to pay attention. It became the same monotony of death. Targets were the same, zigzagging the same, flak, tracer and bullets. Even fear of death was the same.

Sing songs in the Mess cheered him up no end, as long as he didn't overdo the Scotch, but rousing choruses of the Red Flag might not have been most appropriate of songs. It caused whispers and knowing looks among those who reported back to the powers for favours.

Next op the crew came direct from drool-filled sleep. As they gathered at their plane Bill rubbed his bloodshot eyes and watched his breath make clouds of vapour under the lights. He couldn't help but wonder, what the fuck am I doing here?

Over the target flak was so intense at one stage Bill doubted any planes would get through. *Fuck!* It's like I'm flying through an impenetrable wall. He felt his Lanc being buffeted. It kept jumping from side to side with the force of explosions.

If we do get through, we can look forward to doing this again and again, until Jerry kills us.

After their bomb drop while waiting for their camera flash, they became caught and then coned in three massive searchlight beams. The inside of their cockpit was like daylight. *Bloody hell!* Bill looked side-ways at Joe. This was serious. Something needed to be done and quickly, but what? Joe nodded back to Bill with shared concerns.

After their camera flashed, Bill leaned forward. He tapped the altimeter and signed down to Joe who enthusiastically nodded back at him. Bill pushed forward on the stick and pulled sharply away to port. Engine noise pitch changed as he dived. Ben would need to work out a new course. No longer were they part of one massive synchronised metal ballet, nor did they want to be bullet or shrapnel magnets. Their problem now would be if another Lanc dropped bombs from above onto them.

Away from the main stream and out of coned searchlights he felt safer. He set different mental parameters and Ben plotted a new course for their flight home.

While Bill sipped his hot sweet tea another hand on his shoulder indicated a potential problem. No one could raise Skinny. they'd tried. 'Skinny, come in Skinny.'

But from their big tail gunner Skinny, not a sound.

Bill grew increasingly concerned. Intercom failure was common but what if...

He liked Skinny, and thoughts of him in his turret in pain, hurt or dying even from some awful wound inflicted over the target or without oxygen, unsettled him.

He decided he had to know. He must go and check. Bill handed controls over to Joe and indicated he was going back to check on Skinny.

Not an easy task as he crawled through the fuselage out to the rear gun turret.

What with the vibration and numbing cold, it was windy and scary. Using his elbows as front feet Bill squeezed past Skinny's parachute hung on its hook and thought, oh well, at least he hasn't baled out. Not likely to leap without his chute.

Bill found Skinny in his turret. He reached out to check if he was hurt, perhaps dead. Skinny's head was drooped.

'Fuck! He's been hit,' muttered Bill. 'The poor bastard's drowning in his own blood.'

Bill shook Skinny. His heart sank. He felt blood drain from his face. His mouth went dry. A startling revelation; Skinny wasn't in his death throes. Skinny was snoring! Far from being injured or dead, Skinny was dead all right, dead to the world as in fast asleep, slumped over his guns. Bill was flabbergasted. Then he grew angry. Skinny awoke with a start, surprised too, to see Bill. He felt shocked anyone had crawled out to check on him.

'Bill?' shouted Skinny.

'I came to check if you're all right,' yelled Bill.

'I'm sorry, Bill. I fell asleep.'

Having established Skinny was not injured; hurt or dead Bill became angrier.

'Skinny, you're a fucking arsehole!'

Bill crawled back to his cockpit. Normally euphoric at thoughts of returning home Bill pondered this new revelation.

Short of Skinny sitting on Bill's lap to take a crap, not much else would have surprised Bill. But was this normal for Skinny? Did he enjoy a regular snooze when he was supposed to be guarding their rear? Or was it a one off catatonic state?

Not much use a tail gunner fast asleep. When he moved his turret about, at least he looked alert. German pilots might be less likely to attack a Lancaster with its rear guns rotating about up and down.

Not sure what to do, he decided nothing was the answer until at least after they returned home, if in fact they made it back.

The rest of their flight was as uneventful as in peacetime, without a shot being fired. No surprises, no signs of enemy fighters, only the welcome cliffs of Dover.

Bill said nothing of the incident about Skinny at de-briefing, nor did he put a note in his logbook, which at times was vague. Orders were not to detail where they went or what they did.

'Aye. Good luck with that. Makes keeping a bloody logbook summat superfluous,' laughed Olly.

In cold light of dawn Dispersal stood almost empty as so few had made it back. In the Mess their issues were discussed; Joe chaired, Bill remained silent most of the time, while the rest agreed wholeheartedly, whatever actions Ballsy Bill had instigated—good on him.

'We're home! Again, one of the few to make it back,' said Sandy.

But Bill barely listened. He felt tense as he looked about for faces he recognised.

He noted someone familiar was no longer with them, another round of drinks, another comment.

'Aye, attributable if naught else to Ballsy Bill's actions outside protocol,' said Ben. 'Shush,' said Frank, 'You know walls have ears.'

It was unusual for Frank to add much to any conversation. A quiet, nuggetty Englishman by most standards, Bill often wondered if Frank was really a participant at all. About the only thing Bill liked about him was his abnormality. He had a serious binge-drink problem but a handshake like a vicar. He spoke in a rich Midlands brogue and when he walked, it was as if he'd shit himself. Inference was he kicked with the other foot. Geordie ponce!

'Fuck-protocol,' said Skinny. 'We're home and we're safe. We dropped our bombs on their target. Bill saved their plane to fly again. I say if they don't like it let them go fuck-themselves. Word is eight squadrons ceased to exist last night!'

Skinny was right but that didn't prevent Bill from interrupting. He'd decided to broach Skinny's sleep-fest on their flight home.

When he'd finished everyone was horrified. Skinny came under some pretty diabolical fire from his comrades-in-arms.

'Aye, you're a fuckin' arse-hole, laddie,' said Ben. 'I'm doing my bit and you're supposed to watch my wee tail. How can you do that if you're naught but bloody asleep?'

Skinny was not popular with the rest of the crew. He coloured and became angry.

'That's a fuckin' lie,' he argued. 'Bill's wrong. I was awake and alert. Not asleep. On my honour, no, no, no!'

'Sad to say, we might need more than your word,' said Joe.

Mack was amazed. 'How you could sleep back there, knowing the risks we run is beyond me,' he said. 'You're supposed to watch our backs.'

Skinny looked panic-stricken. 'Even if I saw a German fighter, I'm unlikely to fire at him.'

A worthwhile debate followed about whether or not to open fire on an enemy fighter.

Bill was sceptical. 'Look, I'm a bomber pilot, not a fighter pilot. My job is to get us there so Sandy can drop our bloody bombs and get us back again.'

'But, Skinny's right, Bill,' said Joe. 'There's no point in getting tangled up with Jerry, especially when we don't have a snowball's chance in hell of survival. If the German fucker hasn't seen us, we're better not to stir a hornets' nest by announcing ourselves.'

'Agreed. Let him go and leave well enough alone,' said Sandy. 'There's no point in trying to turn a Lanc into a fuckin' Spitfire, because it's not and never will be.'

There was more than a little derision as to what to do to stay alive and what to do about Skinny?

'Aye, but what's the point in a gunner at all, if he never fires his guns?' asked Ben.

'Look,' said Joe, 'I suggest we give Skinny benefit of the doubt and warn him.'

'What?' said Bill, 'Don't go to sleep again!' Bill had kept a smile on his face, but now the edge in his voice gave his true feelings away.

Joe was embarrassed. 'Well, something like that.'

'So, who's going back to check on him to see if he's awake or asleep? Count your fucking peanuts because it isn't going to be me,' ranted Bill. 'You can tell him off if you want. I've had enough.'

Skinny looked excited, as if someone had bought him a beer.

'You would have to have taken your chute off to get through, Bill,' said Joe.

'Certainly did, Joe. No room otherwise. I had trouble squeezing past his chute on the hook.'

Mack looked serious and started to fondle his glass. Airmen at other tables had turned to look at the crew who seemed on the verge of a quarrel.

Then turned their heads away quickly, as if pretending they were unaware of the encounter.

'What's up with you, Mack?' asked Joe.

'We'll never know how many pilots have baled out and left their plane to fly on, with the rest of their crew in ignorance.'

'You serious?' asked Joe. 'Bill and I would never do that.'

'It might not be deliberate. Perhaps they call mayday or tell their crew they're baling out, but no one hears them,' said Mack.

'Poor bastards,' said Joe. 'They'd never know.'

They started to laugh and Skinny joined in, grateful for the diversion. Stories were rife on Base of Lancaster bomber pilots who lost their nerve and baled out, often without due care and attention to their crews.

'That's the bloody worry,' said Mack. 'Not knowing. We haven't a bloody clue what's going on most of the time.' Then he turned to Skinny. 'You! You lying bastard, falling asleep like that. How could anyone sleep on a bloody op?'

Skinny's lies to maintain his innocence might have convinced a few.

'Are you sure he was asleep, Bill?' asked Joe.

'He admitted it back then. He's not quite in the same pond as the other ducks yet.'

There was a titter from Frank then silence you could sever with an axe. Any semblance of togetherness and loyalties were strained beyond belief.

Bill felt guilty. In hindsight he'd felt right to speak out but now he needed somehow to get things back on track. He had an idea. 'Tell you what,' he said. 'Irrespective of Mr Dopey going sleepy byes, how's about I pledge never to bale out without you?'

The group were thoughtful.

'You say that now, Bill. But in the heat of the moment...' said Mack.

Bill looked serious. 'I mean it, Mack.' He gazed one by one at the group around their table. 'I'll pledge it and I mean it. I'll not wear my chute from now on. It's uncomfortable and I'd never have the guts to jump anyway, so I won't wear it.'

The group were moved but sceptical. They were hard to please and easy to offend.

'You serious?' said Mack. 'Not wearing your chute's against regulations, you know. If we took a hit there might be no way, no time, to find it and put it on.'

'I know and I don't care. If they get stroppy I'll sit on it, but I won't strap it on. Bugger it. I'll go down with the plane.'

Joe broke the silence. 'If anything happens to Bill,' he said, 'I think at a pinch I could fly you home.'

They looked at each other.

'If Bill's unable to fly the plane, Joe, what you'll do is hold her steady at about 10,000 feet while we bale out!' said Frank.

'Aye, preferably somewhere over fuckin' England,' added Ben.

'Then you can do what the hell you like with it, Joe,' said Sandy.

'I'll stay with you, Joe,' laughed Bill.

Bill kept his word. He never wore his parachute on an op after that. Often it was no-where near him. It hung on a hook, lay on a rack or he sat on it. Bill's terror of baling out had helped his decision.

Everyone remained in disharmony about Skinny. They were uneasy. Question remained. What should they do about him?

Let him who is without sin, cast the first stone. (Gospel of John)

Chapter 25

FRIENDLY FIRE

The mood of their meeting had been anti-Skinny despite his protestations of innocence to Bill's accusations. This posed a problem for Bill as skipper. What to do?

Joe offered Bill a Fry's Five Boys milk chocolate bar. They were popular at one penny each, when you could get them.

'Thanks,' said Bill. 'Fry's factory in Bristol can't have been hit yet?'

'Not yet,' replied Joe. 'Do you know in 1937 they sold 40 million bars? Not bad aye?'

Bill munched his chocolate and shared his thoughts with Joe. 'If I refuse to fly with Skinny on future ops they'll want to know why. If I tell them, Skinny will almost certainly be stood down, discharged, or even court-martialled.'

'He might be shot,' said Joe. 'Have you considered that?'

All Bill's crew were unhappy about Skinny. No one wanted to fly with him again. But no one was keen to put the boot in and have him harshly punished.

'What should we do?' asked Bill. It was time for a rethink.

Joe reappeared with a tray of glasses and a bottle of whisky. Sandy lit up a Woodbine. Everyone settled down with an air of formality to discuss their situation before they opened the bottle. They were as one. Skinny had to go.

Later an angry Skinny came to see Bill. 'About yesterday, sweet Bill Honey, you suddenly got a problem with me?'

'I wouldn't say that. It's not sudden at all. You might as well have sat at your guns with both thumbs up your arse. I was unhappy with you then, and I'm not any happier now. Nor are the rest of the crew. And while we're at it, you can drop the attitude and the sweet Honey crap or you'll cop a bunch of fives!'

But unbeknown to Skinny, Bill's fervour for fisticuffs had waned.

Skinny sloped off. Later that day he went to the CO. Faced with disgrace of being seen to let his crew down he requested a transfer to another flight crew.

When asked why, he told the officer he was unhappy with Bill's anti-protocol low flying methods, as he believed Bill to be an unsafe pilot with His Majesty's planes.

'Arsehole crawling little bastard!' raged Bill.

The transfer was approved but Bill was sure he had another black mark against him.

The CO called Bill in and had a go at him. He didn't approve of low level flying because of numbers of difficulties; map reading at high speed and minimum height requirements for dropping bombs.

'What did you think about the last op?' It was as if the CO was probing for an attack against Skinny.

'Besides carrying an excessive load of bombs, nothing unusual, Sir.'

'Then your answer should be, about what I would have expected from the style of target, Sir,' fumed the CO.

'Good choice of words, Sir. I'm fighting a war while you're sat here drinking tea and playing at politics. Well done!'

Bill reasoned the attempt at a rollicking was because senior RAF staff was paranoid to protect their reputations at all costs and that wasn't all. Because so many planes were being lost Air Ministry had put out another press statement.

They had invented what they called a Scarecrow Rocket; something they argued was being fired by the Germans to imitate an exploding bomber. The directive claimed the Scarecrow Rocket was a German bluff, and crews had nothing to worry about.

Bill believed it was a lie.

'Aye, now exploding bombers are naught but fireworks.'

Joe agreed.

'So much for trying to help Skinny,' muttered Bill.

'Aye, now you're under their spotlight, Bill, by gum. Naught but a flag or summat against your file for sure.'

'Wouldn't be the first time, Olly.'

In their Mess Bill was more solemn than usual. Joe looked at him. 'You're so quiet. Are you medicated, Bill?'

'I'm angry with Skinny. He acted as if he didn't know whether he'd been shot, fucked or powder burnt.'

Another propaganda bullshit release referenced revival of Clark Gable and Vivien Lee American Civil War epic Gone with the Wind. The Daily Express proclaimed, *The Germans are 20-minute raiders as in—gone with the wind.*

'Does anyone else but us have any idea what's happening?' asked Bill.

Morale of his crew was at rock bottom despite the propaganda bullshit but they weren't the only ones. Chances of surviving a tour were greatly reduced. They were now lower than one in two with anticipated loss of crews and constant threat of death or misadventure. It was considered lunacy to penetrate deep into Germany in any degree of moonlight. And now they faced a new unknown addition to their crew.

CO's remained stiff upper lip. It was as if they disliked anyone who wasn't of high birth. Most of all this particular officer disliked blacks, Irish and colonials so what better than for Skinny's replacement to be an Australian named Barry. He hailed from Melbourne and had completed a short gunnery course in New South Wales before he arrived in England. His accent was a funny blend of cultured English tainted with new world Australian twang.

'G'day mates, call me Bazza for short,' he said, with more than a tinge of an Aussie accent. He had a nugget build and appeared to be highly alert.

He'll be a perfect tail gunner, thought Bill.

'Aye, Bazza you say, naught much different to Barry, if you ask me,' said Ben.

He didn't like the CO. He didn't like flying with colonials and he was barely tolerable of Mack.

Bazza laughed. 'I didn't ask you. Besides I don't give a rat's arse, mate. Like most Aussies I've got a healthy fuckin' disrespect for authority and I'll not likely be fazed by an officer who's got his head rammed up his own arse nor some obnoxious little Scottish git like you.'

'At least you called him Scottish and not Scotch,' observed Sandy.

'What's the difference?' asked Bazza.

'Scotch is a drink,' seethed Ben.

'Oh well, it's off to a good flying start,' said Bill.

'Why don't you like colonials, Ben?' asked Joe.

'Too bloody up themselves and uncultured,' muttered Ben, 'Aye, too brash, too loud. Give me a quiet Englishman anytime if I canna have a wee Scotsman.'

'Bullshit! In a pig's arse, dick-head, there's nothing fucking uncultured about me,' bellowed Bazza.

They all laughed including Ben. 'It's canna like talking to a laddie with the mental retention capacity of soggy newspaper.'

'Skinny was always quiet,' said Joe.

'Yes,' said Bill. 'Quiet all right, except when he snored.'

'Pity more pricks can't learn to be quiet.' Ben looked accusingly at Bazza.

'He's having a go at me,' said Mack. 'Puts me in the same category don't you, Ben?'

'If the shoe fits, laddie,' muttered Ben.

'About Skinny and his snoring, it must have been loud to hear him,' said Joe.

Frank piped in. 'Bill didn't have to hear him. He confessed to being asleep.'

'He'd be about as loud as a pit bull terrier being hot waxed and howling for blood,' laughed Bazza. He seemed to know all about Skinny.

'How accurate is your shooting, Bazza?' asked Ben who wanted to change the subject.

'I'd say it's about average, mate. Out of one hundred rounds fired, accuracy bloody nil,' Bazza said, with a straight face.

'Not bad,' said Frank, catching on.

They continued to trade insults until Joe bought another round of drinks. The drinks kept coming. The night rolled on.

Morning came and Bill knew he hadn't slept enough. A handful of rain like gravel had rattled against the windowpanes. Enough to disturb his rest before his nightmares kicked in and now, the wind had dropped and the rain had let-up but he knew his face would ache all day from lack of sleep.

Their next op was to be another doozy with an even more irregular bomb load.

'Fuck!' Bill said to Joe when they did their pre-flight checks. 'Could they not squeeze another fucker in?'

Joe let out a low whistle. 'Our bomb doors cannot be closed, Bill.'

'I see that.'

To make matters worse it was not to be as dark as they liked, which worried all crews. Darkness was their friend. That the powers contemplated ops with a big bright moon amazed everyone except Olly.

'Aye, but then when pressure's on, if we only ever fly with naught a moon, does limit options,' he said.

If he couldn't have total darkness Bill preferred inclement weather.

'It's turning nasty. If it's bad enough we shouldn't have to go,' said Joe.

Light rain was good, heavy rain was better. Teeming rain was the best. Throw in a storm without lightening and their spirits soared. Sunshine and good weather was not their best friend. This day twilight had expired and the sun was well and truly down with some moon.

On take-off the big birdie seemed more sluggish than normal and with bomb doors not properly closed there was an unusual loud wind noise. It disconcerted Bill. He didn't like it one bit.

At 20,000 feet flak was heavy. Bill dropped down to below 18,000 feet. The flak seemed worse. He dropped down to below 16,000 feet. He would go lower if necessary. *Fuck their cameras!*

They panicked when they opened the bomb doors. Were they really open? Not sure? Someone needed to do a visual. Dropping their bombs with doors partly closed would have been disastrous, especially after arming them, but cookies were always armed.

Everything felt different to what they were used to. Mentally it threw Bill completely. He'd become used to a process. When bomb doors opened a red light came on and he felt the rush of air. With an irregular bomb load, the red light was on all the time and he had all these noises and drafts. They were not normal.

'Pity Churchill isn't up here,' muttered Bill to no one in particular. 'If his life was on the line, I bet we'd see some changes for sure.'

After they dropped their bombs, he was never so relieved as to see those bomb doors properly closed. But before Ben could give a course for home, searchlights picked them up. Tracer criss-crossed the sky and several burning Lancs plummeted downwards.

Not long after Bill felt a bang and an engine became irregular. He feathered it when it caught fire. Whether it had been hit by flak or lost coolant and blew up due to lack of servicing he'd no idea. His damaged Lancaster began to lose height. He tried to remain hidden from the ground in whatever cloud cover there was.

'Crew to check for damage,' called Bill. It turned out to be considerable.

Ben reported his oxygen had been cut off and the GEE was wrecked. 'My wee machine is fucked, Bill.'

Mid-upper gunner Frank reported several holes rear of the fuselage. Joe advised a smashed oxygen regulator was leaking. He estimated their oxygen would be depleted within twenty minutes. At their prescribed height that would cause anoxia, a condition likened to drunkenness followed by unconsciousness and death. Bazza said one hole was so big it had taken his chute off the hook.

Bill sweated under the pressure. He wound the altimeter down to under 8,000 feet in a shallow dive without any idea of his exact height. It was a successful attempt to negate the need for oxygen and put out the fire. He planned to limp home at an even lower altitude on three engines. Joe had a constant eye on falling fuel gauges. It was a fragile reminder of how short-term their lives were. He channelled fuel supplies from the stricken engine to the other tanks.

They returned over the North Sea instead of the English Channel and Bill was made aware of an unidentified smaller plane following them.

Bazza's voice sounded strained. 'There's an alert, Bill! We've got something behind us.'

'I see him,' chimed in Frank.

'Terrific,' muttered Bill. 'That's all we need.'

Their technology was crude but rising animated conjecture feared the worst. It could be an enemy trooper's Messerschmitt or Junkers Ju8.

Mack indicated what was left of his board had lit up like a Christmas tree. It appeared impressive with the right amount of lights, bells and whistles to attract every German plane for miles.

Bill sensed some semblance of organised panic on board. He decided to take immediate evasive action but didn't have enough height for the corkscrew manoeuvre.

Joe shook his head to indicate their fuel supply might be a problem. *Fuck!* Instead he changed course and swung to port. Bill hoped if the other plane were a 'friendly' he wouldn't continue to follow. Unfortunately he did.

Had he referred to their same 'manuals' he shouldn't have done that, thought Bill. He should have signalled his 'friendly' intentions and broken away.

Bloody hell!

Bill changed course again, swung back to starboard. But the unidentified plane remained on their tail and was gaining. The radio crackled in Bill's ear and despite the seriousness of their situation, he couldn't help but laugh. It reminded him of Fred's favourite story from his war, the First World

War. The one he'd told to young Eric on his last leave about when radio communication was in its infancy and hard to understand.

As it continued to crackle, Bill wondered how anyone could make sense of it. But if they waited and it wasn't a friendly, it could spell their death warrant.

What with noise and vibration, Wireless Operator Mack couldn't decipher anything at all. He was almost at his wits end.

It seemed to be one of those times when to curl up into the foetal position could become a favoured option. Then upper gunner Frank and rear gunner Bazza both shouted, 'Go, Bill! Dive!'

Bill dived with an eye to his altimeter. As he did so, they opened fire with everything the big old bird had——simultaneously. Noise and vibration shook them like horrendous thunder throughout the plane. It was one of the few times both gunners had fired their guns at the same time.

Gunners Bazza and Frank spoke over the top of each other in their excitement.

'A hit, I claim a hit.'

Bill was sceptical. Undeniably their intention was to hit their target. And with adrenalin pumping, a most welcome surprise when it appeared they had.

Each claimed a visual. 'We got him!'

What they got, they were unsure, but they thought something went down. It veered away and stopped following them. No one saw any signs of a parachute or a pilot baling out despite there being some moon.

The rest of their flight home passed without incident. Bill brought the big plane into land and was surprised how well she handled.

After their return to Base, they had their mandatory de-briefing but not everyone rejoiced as loudly about their boasted news of downing an enemy plane. Their claims of good marksmanship were over-ridden by an intelligence officer with bad news. 'A Spitfire has been reported missing same time and location,' he said.

The assumption swelled their festival of shame to maximum pitch. The plane Bill's crew had shot down was most likely one of their own. A friendly!

An officer had a go at Bill, 'you're not trigger-happy Yanks, so stop behaving like them.'

Bill was put out but conjectured being lost and frightened, perhaps the other pilot saw their lumbering giant and thought it easier to follow them home. He might have been one of those fresh-faced young pilots with less than two weeks training and because he lacked experience, didn't understand why this Lancaster kept changing course. Unaware of the routine expected of him, he might have been injured, or thought following the Lanc was an easier solution to return home. Not unlike another young and inexperienced cadet pilot who had tried to follow a river home in Canada. Bill winced. Another wasted life and a future that wasn't to be.

'Not that any other Lanc crew wouldn't have done the same in those circumstances,' said Joe.

'It's not easy to bale out of a fighter plane,' said Bazza. 'First, the pilot needs to have the presence of mind to turn the bloody plane upside down and physically remove the cockpit cover. Then disconnect his harness before allowing mother gravity to do the rest and let him fall out.'

'Not easy for a fit man, more difficult if injured, and next to impossible if the plane's out of control, nose diving, or burning up at the time,' said Frank.

Bill summed up. 'Trying times,' he said. 'Affecting so many eternally.' But for days Bill's crew were ostracised. General hubbub would cease when they entered a mess. Some stopped talking or looked away. * 9th Footnote

Also at de-briefing—there was no sign of Skinny. Missing in Action, his plane given up as lost.

Chapter 26

GESTAPO HQ

Daily Express placards carried the headline; *96 down!*

Bill felt awful about Skinny. In an effort to placate his conscience he mentioned him falling asleep to Lily and Fred. Fred shook his head in amazement and was supportive, but Lily offered another perspective.

'I'm not surprised,' she said.

'Why, Mum?'

'Well, it's a bit like being a child, isn't it,' explained Lily. 'When a child is upset or scared they often suck their thumb and go to sleep. It's a form of escape.'

Bill didn't feel any better after his mum's input. Now he felt guiltier than ever about his handling of Skinny.

After a short leave Bill returned to Base.

Their next op was shrouded in the usual security. Likely to be an open secret at the pub, their target was only disclosed hours before take off when the curtain was raised from the end wall.

'Not that it will make any difference,' grumbled Joe.

'Not wrong,' said Mack. 'Entire German army will be expecting us for sure.'

Their target was announced. Gestapo Headquarters.

'Are they serious?' asked Frank. 'We're actually gonna bomb them right in their own back-yard?'

Bill's face was a mega smile. 'Deep in the heart of their Black Forest!'

'Yahoo!'

Frank lit another Woodbine. Bill looked at him. He held his cigarette effeminately, more like a woman. There was no way Bill felt he would ever hit it off with this man. And there was another thing; their missing shirts. Not only his, but others too. If Frank needed a clean shirt he wasn't averse to helping himself from another man's locker. He never asked permission. No by your leave, nothing. He simply took a clean shirt if he fancied it. That didn't go down well with the rest of the crew.

Back to their target, dedicated as military, it was one that supported the struggle for Allied crews and foot soldiers alike. Bill was speechless with admiration for his superiors.

'At last they've done something we can be proud of,' he said.

'Yes,' said Joe. 'We'll hit the bastard's hard, right where it hurts them most.'

This was not any target but one to be savoured and enjoyed. Gestapo were about as popular with Allied Forces as foxes loose in a chicken coop. More interest than usual was taken in briefing. There were less complaints and not a single concern about the irregular size of bomb loads.

They were told to expect thick cloud and a head wind of 25 mph.

Unlikely, thought Bill. The stronger the wind less likely thick cloud.

Surely everyone knew that.

Planes set off more enthusiastically than usual even with bomb doors locked open. Bill felt as if their undercarriage was being dragged along the ground.

'We're loaded just short of our undercarriage buckling,' said Bill.

'Aye, but they don't have to fly it,' said Ben.

True. If an aircraft failed to lift and crashed, killing the crew, it wasn't the powers' lives at risk.

Some Lancs carried the new twenty-two ton Grand Slam bombs. Others had the twelve tons Tall Boy bomb. But everyone was in considerably better humour than usual because of the popularity of target. Their only downer had been a watch on the weather. Light rain had started to spit from the grey sky. Then the wind increased. The odd rumble from the east indicated a storm might be brewing. No one wanted this op called off. Nothing would have dampened their enthusiasm.

In the otherwise dead of night, a moon hung above bugger-all cloud. There was only a light tailwind. Fuck me, thought Bill, could they have got this more wrong?

Over the target visibility was clear. Then flack started. Bill looked sideways in the gloom at Joe, as he often did. They both nodded to each other as if to say, it's dark down there all right, but that's because we're over the Black Forest. Right?

Bill's plane was frontal of the attack, which meant most action would be behind him.

'Bill, turn the wee plane five degrees left.' Then the gloved hand on his shoulder indicated they were on course and ready for their bomb run. Bomb doors were opened.

There was a short wait.

'Steady, steady—bombs gone!' Then that distinctive surge upwards followed by, 'Bomb doors closed.'

Seconds ticked by, 'Hold, hold, hold—wait for those pictures.'

Bill glanced at his altimeter. Pity he couldn't rely on it. More time ticked by.

Great sheets of light below lit up the sky. A second touch on his shoulder along with those words he longed to hear. 'Time to go home.'

'Shouldn't be too many tall buildings jutting up out of the Black Forest,' said Bill to himself. 'Let's give the wild-life a hurry up.'

As Bill forced the nose of his big plane downwards he noted his increase in speed. Pity we can't go this fast normally, he thought.

From the corner of his eye he caught a flash and a shadow as a fighter plane cut across, Bill didn't think he'd seen them. He hoped he hadn't.

He increased the angle of his dive. If he wants me, he'll have his work cut out.

Suddenly there was one hell of an explosion. The entire plane shook. It was if a giant hand had reached out and pummelled them. If he hadn't been flying at the height he was he would have sworn he'd hit a tree!

Bill and Joe were showered with metal fragments and each flinched as if expecting to be blown to smithereens. Tentatively Bill leaned back on the stick and was surprised when the old girl responded. He looked down at his legs and was pleased to see them still there.

What the fuck was that?

'Crew, damage report.' He cast an anxious look around. They appeared to be in one piece although something was different. He looked across at Joe who gestured upwards.

Bill looked up to see sky where there shouldn't be any. He felt a strong draft about his legs. It confirmed he had feelings in them. It was hard to figure, as he struggled to come to terms with having been hit, he levelled out to maintain his course. Only then did he start to piece their emergency together.

They had been hit all right. The one sound a pilot never wants to hear.

Most likely a mortar shell had exploded nearby. Not a direct hit, or they would have been history. And no amount of training fully prepares for that, no matter what. What might have been shrapnel had passed right through the plane. It had come in through the floor, missed him and Joe by a few feet, and burst its way out through their cabin roof to their starboard side.

The force had been incredible. No wonder the plane shuddered. Around the inside, everything appeared as in a dense brown fog. It slowly gave way to a haze, then to clean crisp air. Dust and debris were everywhere. Bill had been showered in it, and now wind rushed in where none was supposed to be. Unsecured equipment and anything unattached was airborne. He was in the middle of what looked like a dusty version of Dante's Inferno. Screams mingled with shouted commands as Bill checked his instrumentation. He tried to see if everything was working. Wind howled and whistled. It bored through Bill's brain faster than he could ever imagine. Joe put out a small fire with an extinguisher. No mean feat. He gave Bill a hard glance.

'I don't see any blood,' shouted Bill.

One engine was starved of fuel. There was a strange noise and when Bill looked he saw another engine had caught fire. Flames licked around it. They made a streaky yellow glow where there shouldn't be any. Once feathered the flames diminished and went out.

Fuck! They now had to limp home on two engines. That's twice in a row. Thank Christ the bombs were gone and no hang ups.

Joe tried to check on other crewmembers, from what he could see they had no serious injuries. Ben was huddled over in pain. He felt he might have cracked ribs. Mack was slumped over his wireless equipment. He had torn clothes from light shrapnel wounds and was badly shaken. He nodded his head to signify he was all right.

Everyone appeared to be intact but shaken. Watching fuel gauges drop, Joe closed off the fuel selector to the shut down engine. He kicked himself. He should have done that sooner before checking on Ben and Mack.

In the gloom they all had the same strained and panic-stricken expressions on their faces.

'Just another glorious day at the office,' joked Bill.

It dawned on Bill they were in trouble and it made a fist of anxiety in his guts. Adrenalin rushed through his body and his training clicked in. Don't panic! But he could feel it rise in his throat like hot lava. He knew he had to look and behave normally. But should he give the order to bale out?

'Nothing kills a pilot quicker than panic,' his instructor had said.

He did another instrument check. His eyes raced across the dials. No, he was panicking. It was all too fast. He wasn't taking it in. Slow down, Bill. Get it right. Check but slowly...

It was as if the plane wouldn't fly straight. But it was flying. He did his best to trim the aircraft but the ungainly Lanc had other ideas. As FOB would say, 'It's not renowned for its gliding capabilities, Cadet.'

He needed to compensate for the damage. He signalled to Joe who signalled back. He was all right. He looked about again. No smoke, no fires. Bill tightened his grip on the controls. As he lifted her nose slightly, she slowed.

Good. At least that's normal.

He tipped the nose forward and their air speed increased but the plane began to judder too much for their current speed. Fuck! She didn't like that. Not one bit. He eased up. He'd better not do that again.

He flew level as he could with two engines. What he didn't like was not going straight. It felt as if he was moving through the air, crab like, at an angle. And he needed to maintain altitude.

'Tell everybody. Jettison any weight you can,' shouted Bill. 'Anything you can, haul it out!'

Bishop to G6, thought Bill. Landing her should prove interesting, if we get to land!

Bill dared not breathe. At any moment he feared the plane might do its own thing. It might go into some chaotic manoeuvre of its own accord, out of which he would be unable to coax it. All the way back Bill worried he might lose control of the plane. Maintaining height was a serious problem.

Joe tapped his dials with an anxious watch on fuel as Bill nursed them home.

They crossed the English Channel and miraculously Bill gained height. It would soon be light.

A squad of three Spitfires sped past him. One circled and returned alongside. He wobbled his wings as if to say, you'll be right, Skipper. I'm with you. I've seen you.

At least his radio might work, thought Bill. If he sees us go down he could call in our position. Below them the sea looked sullen, an unwelcome leaden grey. Bill shivered at the prospect of having to ditch. It looked so cold and hostile. He tightened his grip on the controls. Not now, we're so close. Please don't fail me now!

They were almost home before the Spitfire accelerated his incredibly powerful 2,050-hp engine to about 500 mph. Bill watched him vanish into the dawn. As he surged away he thought of Duncan, and how fighter pilots were dying at the rate of five a day.

'Wouldn't mind knowing who that young fellow was,' said Bill to himself. He was dimly aware his bum had gone numb.

Ben called the nearest aerodrome for permission to land.

'Permission denied!'

Joe signalled the hydraulics to lower their undercarriage was not working properly. He clambered out of his seat to check their wheels were locked down and was pleased to find they were. Pumping the manual under carriage lever back and forth to lower or raise the wheels by hand was a tiresome job and would have taken too long.

As they approached their own main runway all hell had broken loose below them. There was smoke and fires and signs of attack everywhere. Lights flashed and people ran in all directions. Now it made sense why the nearest aerodrome to the coast had refused them. Perhaps they were out of action due to a raid.

Bill brought the big bird to within a yard of touchdown. Sweat stung his eyes.

Then he eased her onto the concrete. He hoped like hell the landing gear would hold. With brakes applied and propellers feathered the great plane groaned to a halt.

Emergency trucks and ambulances quickly surrounded them. Someone was sobbing. Bill heard but decided not to find out who it was. Sometimes those things were best ignored.

He looked at Joe, and Joe looked at him. They both sat stunned.

'Well done, Skipper,' said Joe. 'For a moment I thought we might not make it.'

'You and me both, Joe,' His stomach had not stopped doing cartwheels.

After they exited the plane Bill walked around her and whistled at the damage. They had been hit in a number of places. No wonder the Spitfire had noticed them. There were holes the size of footballs where metal hung limp and fabric had flapped about as they flew. Both wings were peppered. The fuselage showed massive holes and who knew what had passed through the floor and out the top of the canopy? But whatever it was had left spaces big enough to party in. Miraculously no one was seriously hurt.

'Can't see this old bird being ready to fly anytime soon,' said Joe.

Bill shook his head, 'You'd be surprised how quickly they'll patch her up.'

Around them were signs of chaos and destruction. 'Let's get out of here,' said Bill.

It made him sick to look at the mess and think about what it might take to put the old girl back into order.

Then the heavens opened up and dropped sheets of water. It rained hard. Potholes filled and became a grey sludge and then produced a rainbow effect; spilled oil and fuel lay everywhere.

Bill picked his way with care. 'Watch your footing,' he said. 'All we need is casualties now we're down.'

Joe was pessimistic. 'This Lanc's definitely one for ground crew to sort out. That is, if even they can get her to fly again.'

'Yes, they might need their magic feather this time,' laughed Bill.

I couldn't fix your brakes, so I made your horn louder...(Anon.)

Chapter 27

FLUKE

Their human loss was greater than usual. Less than half of Bill's squadron had returned despite assurances to the contrary. Jerry had expected them.

Also at de-briefing there was no Tache. Missing in Action, his plane given up as lost.

'Yes, they knew we were coming, all right,' said Joe.

Bill became angry. 'I'd like to catch up with the fucker who keeps telling them,' he said.

'You and me both.'

Bill found himself looking for Olly, and yes he was there. He'd made it back. He saw Bill and gave his cheery wave, then he feigned with his hand over his heart, as if to say he was pleased to see Bill had made it back too.

At de-briefing there was considerable interest. But not until after development of their bomb placement photographs did news of the action begin to unfold.

Complete disaster!

Not one bomb had touched the Gestapo headquarters. Its gaunt, defensive outlines had been more softened by years of rain and wind than their onslaught. There it stood, undamaged, as testament to its chapters of history.

Their disappointment was palpable.

Aircrews were left reeling. They felt more than usual their full burden, the incredible loss of their fellow comrades—and for what? Many became de-humanised at the thought of their direct involvement in what speedily changed from a mission of gravity to one of farce. A massive loss of fellow mates and comrades totalled over 200.

Many of Bill's colleagues, those lucky enough to return never recovered psychologically. That op changed their personalities. In the mess, young crews sat around without interest. Some acted in a stupor, as if they were in another world. There was quiet, and then for no apparent reason, a sudden outburst. Someone would throw their chair back hard against a wall and storm out, or overturn a table of empty glasses. About him young men behaved in strange ways. There was no interest in sing-along's and the piano normally busy, stood silent.

Suddenly the entry doors burst open and there sat Olly astride his Triumph motorbike. He was drunk. As he revved the throttle and dropped the clutch everyone in his path scattered. The bike shot forward until it hit the highly polished floor when his back wheel spun out of control. Instead of whatever entry Olly had in mind the Triumph had different ideas? It lurched sideways and took two hurriedly vacated tables and umpteen chairs. It then deposited him head first into the trophy cabinet. Wearing his flying gear, Olly was well padded, but this was not the accepted Royal Air Force way to enter a Mess.

Bill went over to Olly spread-eagled on the floor. The pallor of his face was tinged with a delicate green. He looked unwell.

'Are you all right, Olly?'

'Aye, Bill, but I've found the perfect fuckin' way to enter their bar.' He slurred his words then erupted like a volcano. His vomit splashed across Bill's shoes.

When Bill was able to look deep into Olly's eyes he read a profound distaste for the whole business, but was there something else there that troubled Olly maybe beyond the war.

'Aye, summat wrong somewhere.' Olly said as he wiped spew from his mouth with his sleeve.

Bill and Joe helped Olly to his billet. The rest of the crews assisted with clean up. Joe returned to the Mess leaving Bill alone with Olly.

'Not like you to make a spectacular entrance, Olly,' said Bill, as he wiped his shoes with Olly's face flannel.

'Aye, I'm naught but at my wits end, Bill, by gum.'

'Why? What's wrong?'

'Maintenance or lack of it. These planes, they're naught fit to fuckin' fly.'

'What in particular troubles you, Olly?'

'Too much is wrong, Bill. Aye, they only patch them up. Last op I called into Dispatch for fuel. I'd swear they forgot to put any in. Time before hydraulics hadn't been bled. They even attached bombs to a rack clearly marked unserviceable.'

Bill was sympathetic. 'It pays to check. If you're sure there's something wrong, you can always refuse to take her up you know. Ultimate responsibility lies with you, the pilot.'

'Aye and you know as well as I do what a fuss that'd create. I threatened not to fly last time because starboard outer leaked streams of glycol. Bloody bastards threatened me with LMF—the turds.'

'Oh, well. Try and get some rest, Olly. I'm sure you'll feel better in the cold light of day.'

Bill left his friend, dry retching into a bucket and returned to his crew. They all regrouped for another drink. Except for shards of broken glass in the doorframe of the trophy cabinet the room appeared as before.

'How was he?' asked Joe.

'He's pretty upset. Claims maintenance on planes is well below par,' answered Bill.

'He's not the only one who thinks that,' chimed Sandy.

'But how could this have happened?' asked Joe.

'What? Olly pissed and riding in on his motorbike?' laughed Bill.

'No, it's not about him. How couldn't one bomb hit Gestapo's headquarters? Were we deliberately given the wrong co-ordinates?'

Bill was thoughtful. 'Could be. I don't trust them. But more likely they fucked up than did anything deliberate. They're not really clever enough to do anything intentional.'

'Aye, but those German lads are canny bastards. Bloody clever,' mused Ben.

Bill turned to bomb-aimer, Sandy and asked, 'How was your run-in?'

Sandy was always a quiet man without much to say. His eyes were dull as pennies.

'Perfect, Bill, couldn't have gone better. I input details of bomb load, target altitude and wind direction. There was next to no wind over the target. Clear in the cross hairs. I thought we were home and hosed.'

'How could calculations be that far out?' asked Joe.

'Was it the co-ordinates?' offered Frank.

'Stronger than expected winds perhaps?' suggested Mack.

'Did we go in too high lads?' asked Ben.

'No,' said Sandy.

'Not much chance of that,' said Joe, 'not with Ballsy Bill at the helm.'

Sandy was thoughtful. 'If our height was wrong, we'd likely be too low, not too high.'

'Was it our navigation?' asked Frank. He looked directly at Ben.

'Or was it plain bad luck?' offered Bill to defuse the blame game.

'Doesn't take into account bombs dropped over the sea or what they can't account for,' said Bill. 'It sifts only the information it wants, to show how inaccurate our bombing is.' * 7th Footnote

Sandy frowned. 'Their report maintains less than one third of bombs dropped were within five miles radius of the target.'

'We didn't do too badly then,' Joe said sarcastically.

'Depends how you define the word target,' argued Sandy.

'Trouble is the Butt Report condemns accuracy of night time bombing,' said Frank.

'So what's your point?' asked Joe.

Sandy looked serious. 'Well, if night time bombing is proven to be too inaccurate, rather than fix the problem, their easy out is to send us—in daylight.'

Joe exploded. 'Fuck! I don't fancy that. Bad enough now but we'd go down in bloody droves.'

Uneasy nerves pulled at everyone's guts. No one liked too much of a mystery, particularly when it was their own necks on the block.

Base had a priest, Pastor Alistair. He was a well meaning, gentle and devout man. Whenever he was around, for some, heaven seemed a little nearer.

'Few of my youthful dreams of a life spent helping others in remote parts of the world have been realised,' he said in his usual soft voice.

By comparison crews around him seemed to bellow.

'You can only do your best, Father,' said Joe. 'Better than being a recruiter for the other side.'

They laughed.

'Personally these days I find the answer first in the Lord, then in strong drink,' said Pastor Alistair.

He was a man who at his pastoral best seemed safe within his convictions.

'Join us then, Alistair,' offered Bill.

'Yes, I will. Thank you. Everything does seem funnier after a few beers, doesn't it? After-all death comes to us all.'

If that's the best you can get from a priest, I could do the job, thought Bill.

Alistair was a strange mixture of religion. He seemed willing to be guided by the judgment of others about most New Testament miracles. As a welcome diversion, Joe asked his opinion about the 10,000 who came to hear Jesus and the parable of the loaves and fishes.

'Perhaps some had provisions hidden within their robes,' smiled Alistair. 'And they were sufficiently moved by the message of love and brotherhood from Jesus, to share an otherwise privately hoarded sustenance.'

'More an early miracle of communism, then,' observed Joe, 'than a multiplication of food.'

'Any miracle will do me,' said Bill. 'Might He part the hostile skies for us tomorrow, too, Father?'

Night came and Bill struggled to fall asleep. Outside was sleet and darker flakes had fallen. Despite being overtired he found sleep difficult. His mind continually needed to unfold the day's events and his dreams, again became nightmares. His Lanc was holed by flak. This time he had no controls. His flaps and hydraulics were shot away. He must surely crash. Oxygen tanks exploded. Flames spread throughout his Lanc. It was worse than terrible. Horrific! He went looking for Richard to help him with his parachute, but he wasn't there. Olly was there instead.

In his dream he helped Olly to put on his parachute to bale out. He watched another Lanc go down. Again he saw Skinny in the rear gun turret, his sobs appeared as tracer. His eye sockets were guns. That's a clever trick, thought Bill. I wonder how he did that?

As he fell down into the abyss Skinny reached out both arms to Bill and shouted, 'I didn't fall asleep, Bill. Honest I didn't.' And he was gone.

Bill awoke in a cold sweat. *Fuck!* He lay there afraid to sleep in case his demons revisited him. He felt as if he'd been connected to an improperly calibrated drip of adrenalin.

Soon enhanced bomb placement pictures showed the full sorry story. The forest around the fortress had been completely annihilated; hardly a tree was left standing.

'Pity the poor wildlife,' grumbled Joe.

'Those deer wouldn't have known what hit them,' said Frank.

'Aye, we gave those German bastards a wee roast venison dinner without even trying, lads,' laughed Ben.

Of the main buildings not a stone or brick was dislodged.

Frank shook his head. 'Only some areas might have been slightly scorched or singed, then,'

'What a complete waste of time and lives,' said Joe.

For about a week the open feeling of hopelessness and gloom continued throughout the Squadron. Their moods shifted slowly through the full emotional roller-coaster ride.

Then something amazing happened. News trickled in from French Résistance in the area. The bad news was, yes, there had been the usual but rarely acknowledged security leak and the Gestapo had been expecting the massive raid.

'That explains the heavy flak and our appalling casualties,' said Joe.

'No surprises there,' said Bill.

The more positive news was the Gestapo had vacated their headquarters prior to the impending attack and taken shelter in the forest. They even took most of their secret files and records with them for safety. The op by fluke of circumstances had been a complete success after all.

Mack crossed himself. Sandy lit up a Woodbine.

'Lord have mercy on our children,' offered Joe.

Bill leant his head back and tipped the last of his Scotch down his throat. 'As it's doubtful any other fucker will,' he said.

Cry havoc, and let slip the dogs of war... (William Shakespeare)

Chapter 28

ALICE

Winter 1944 was drawing to a close. The usual London rain drummed against their windows. When it decreased to a cheerless drizzle, Bomber Harris called on an average of 1,000 bombers a night to fulfil his personal ambitions. Even more would be provided to clear the way for D-Day and destruction of the Reich.

His tour well underway, blue eyes the shade of dishwater, Bill was left with his memories of a winter campaign, which had sucked. It had claimed the lives or freedom of nearly 6,000 fellow aircrews since his first flight with Tache.

Now as scent of summer flooded the dim afternoon, Bomber Command had a new tactic, one to cause considerable resentment in every squadron. Because more and more raids involved ops over French and Belgian targets, the RAF announced these were to be counted as only one third of an op.

This innovation meant increasing the length of a tour from thirty ops. Such an edict meant an even greater chance of not surviving the conflict and caused further unrest among bomber crews. In the Mess was open talk of mutiny, which grew after their savage losses in the op over Mailly-le-Camp in France. Designated an easy French target it had counted less than a normal op.

'It shouldn't mean us losing over fifty Lancasters,' complained Joe.

'Bloody Churchill and his war. That's 350 brave men, gone,' ranted Bill.

Their situation had not been helped by further self propelled ideas of brilliance by Bomber Command in trying to improve on Wing Commander Guy Gibson's use of VHF radio, when he'd guided other Lancs onto dam wall targets. Developed to a point of stupidity, their main wave was to orbit above a coloured marker detonated at about a thousand feet and then wait around before being called in to drop their bombs. Men unwisely educated beyond their intelligence intended to improve bombing accuracy, but their plan backfired. Instead it resulted in an open invitation to slaughter by German fighters.

'Why don't we send the Luftwaffe a bloody open invitation?'

'Aye, Bill, we might as well be draped in naught but tricolours,' groaned Ben.

Bill was horrified. If ever he and his crew needed proof anything orchestrated by the RAF had cock-up written all over it, this was a prime example. His taut face was as hard as cement. 'No wonder so many good men go AWOL. What are those idiots at Bomber Command thinking?'

'That's the trouble,' said Bill, 'they're not thinking. It's a clusterfuck.'

The fight had left him, but not the rage. It boiled behind his eyes and seared red-tinged images into his brain.

Bill began to realise even if he was promoted, no one in Bomber Command was ever likely to listen to anything he had to offer. Disillusioned and angry he switched off mentally and spent whatever time he could, when he wasn't engaged in killing for His Majesty, at home in Barlby Road. Where possible he tried to distance himself from the bureaucratic buffoonery. He tried to tune out the clamour around him.

But his thoughts often returned to Sophie. They'd never kept in touch as she'd wanted and Bill couldn't rid her completely from his mind. He wondered if her husband had ever returned to her. And when he had those thoughts he felt guilty.

After winning combat against his conscience, he sat with a blank piece of paper in front of him. He meant to write to Sophie.

A few times he put pen to paper, but after a clumsy sentence or two, he gave it away as a bad job. He'd started to write:

Dear Sophie,

I hope this letter finds you happy and well, and your husband has returned from war...

He stopped. What now? Do I write; if your husband or someone else isn't shagging the arse off you, I sure would like to give it another go!

They had been as intimate with each other as playing cards in a pack. How could he ever write that style of letter? What if her husband had returned and they were together? Happy? Bill realised he lacked the skills to write such a letter and his inability to do so further angered him. But whatever the content, if he wrote it, it would be signed by him, and the last thing Sophie might want now, was a reminder of him.

He took a break, went to the window and from there he saw a half moon, and as he watched, as if disgraced, he saw it trail through rags of ominous black clouds.

An omen perhaps.

Time after time, Bill tried to write his letter to Sophie only to fuel the fragments into the raging hob. He felt embarrassed should anyone see his pitiful attempt to renew his friendship with a married woman, far, far away. And what if Sophie responded favourably?

What if, for some unlikely reason, she decided to resurrect their association? What then was he to do? He could hardly pack up his kit bag and hop a ship to Canada. It was so far away, it might as well be the moon.

Bill was never to see Sophie again, nor did he ever write to her, but he had difficulty getting the Canadian beauty out of his mind. He grew more and more depressed.

With his expression as stony as the wall of a medieval town under siege, in desperation he put a few pounds in his pocket and went for a stroll down Bayswater Road, Kensington.

Known locally as prostitute mile, along the kerbside of the northern end of Kensington Gardens women of all ages stood, almost shoulder to shoulder waiting to make amiable terms.

The place crawled with mostly polite, frightened American boys. They didn't find it easy to socialise with the majority of British people, who had a healthy disregard for them. Resentment of their smarter uniforms and higher rates of pay played a major role within British forces. Bill looked a few US airmen up and down. They even had better sets of metal combat wings. The badge of the US Air force was beautifully crafted, ostentatious compared with his cloth ones. And their Golden Eagle, as it was called, always shit plenty on Fridays, which meant these boys had money to burn.

Bill shouldered his way between dozens of Yanks of all ranks who wanted to get laid. The GI's bartered with US dollars. It's as if they all live the American dream, he thought; plenty of money, endless supplies of cigarettes. Treats like nylons and chocolates for any woman out at night without a husband. Even the poorest had more money to spend than he had. He realised no matter how hard he worked he'd never match them.

About 300 prostitutes stood in a long line, as each tried to attract attention of anyone with a wallet and testicles. Time after time Bill was accosted.

'Want a naughty girl, Mister?'

'Only cost a pound, Love. Any way you like it.'

'I'll give you a suck and a shag, Sweetheart and all for only ten bob. Best rate in town.'

Bill winced.

He looked a few up and down. But there was nothing about them he liked. Their cheap bracelets and gaudy jewellery clattered and jarred Bill's mind. Lipstick in slashes of harsh dramatic reds was painted on their faces like some sort of badge of servitude. With some it was as if their throats had been slit at a point under their noses. Their eye shadow applied heavier than any serving girl from last century, rolled into fresh hairline rills about their eyes.

Compared with the attractive women he'd known; Sophie, Anne, and Jill, these women looked rough and held no appeal for Bill.

Perhaps the older a bloke gets, the lovelier women appear to be, thought Bill. In which case if I make seventy, there might be advantages. Every drab one of these will be as a golden fucking jewel.

And yet they were busy. The Yanks couldn't get enough of them. No sooner had one left arm in arm than another appeared and took her space.

To Bill their heavy make-up and crass approach to their craft lacked any finesse. Their gin sodden breaths and addiction to cigarettes did nothing to attract him. His normally erect and healthy penis shrivelled-up at the thought of them touching it, let alone him putting it inside, even protected with a condom. He felt about as virile as when he'd tried to penetrate Sophie's plaster cast.

As a further badge of servitude these women wore gaudy anklets and with a shudder Bill acknowledged Bayswater Road wasn't for him. He broke away to return home.

He caught a Central tube train from Queensway Station opposite, then hopped off and boarded a bus. He needed time to think.

Oversized raindrops plopped onto the pavement followed by another week of grey skies and nonstop drizzle. It kept the gardeners happy but made Bill feel morose. Then the weather changed to an unusual sky, which brought blustery conditions.

Bill left his mum and headed out to visit one of his cousins who lived in Acklam Road, a fair walk from where he was. He looked down at the polished blunt toes of his shoes protruding from under his uniform trousers. Not only had the RAF taught him to fly but they'd taught him to march. A few miles were unlikely to bother him. He shouldered his collar against the stiff breeze and set off, left right, left right.

As Bill strode past Ladbroke Grove tube station he realised he was about to enter his old cricket bat territory. But now, somehow, he felt different.

'I've killed a few thousand Germans since I was last here,' Bill laughed quietly to himself. 'And with any luck I'll kill a few thousand more before they kill me.'

A tinge of green had appeared in the sky to the east. I'm pleased I'm not flying tonight, he thought. Then again they might get lucky and be stood down.

Bill would never know what circumstances conspired in his favour that day. Perhaps it was a combination of things. Not being under the influence of tea laced with bromide for the past forty-eight hours probably helped. Feeling confident in his smart blue uniform might have contributed, and proudly sporting his wings, rank and badges lent a hand. It was his old neighbourhood but he no longer felt threatened.

Then he saw Alf.

He'd been taken aback by the change in people. It was as if he'd left them young and in only a few years they'd softened into maturity. The plump were plumper but Alf didn't appear nearly as big as he remembered and now he wore a dustman's clothes. The great leather pads on his shoulders and across his back were well placed to protect him from heavy metal bins he lifted and carried. He looked as he'd always looked, as if he could knock Bill down, and might decide to do it before he wiped his feet on him.

Stan was with him, but there was no sign of Jack. Although these were the school-yard bullies Bill had feared less than ten years ago he felt at ease, even without his trusty bat.

Stan removed a hat Scruffy wouldn't have chewed on and nodded. Alf's grimy fingers touched briefly the peak of his soft cap. Neither spoke, words seemed unnecessary. Their approval of Bill's rank and uniform showed in the lowering of their eyes and their embarrassment by the way they shuffled their feet.

Bill recognised their respect for his uniform or him, he was unsure, perhaps a bit of both. But Alf and Stan didn't see scrawny ineffectual Bill, either. Before them was a qualified RAF pilot, his insignia and what he stood for, his country, and their freedom clear to see. What they couldn't see was his inner turmoil. The enormous loss of so many comrades and the trauma associated with trying to stay alive.

Bill spoke, 'Where's Jack?'

Stan looked uncomfortable. He stared at the pavement. 'He's joined up,' he mumbled. 'He's a stoker in the Merchant Navy. We're waiting to hear when we'll be called up.'

'Not volunteering then? Not signing up until you're called?' asked Bill.

He heard briskness and aggression in his own voice and felt shocked and slightly ashamed at the sound of it. Asserting his authority was hardly necessary, not now.

Alf and Stan didn't reply. They looked even more uncomfortable and shook their heads. Then without a word they doffed their caps and turned away.

His uniform had bought him additional respect from kids he'd been to school with. They understood what pilots were and how few survived. Even allowing for lies and propaganda, news trickled through of heavy losses, which elevated Bill to a status earned by a rare few.

After Alf and Stan wandered off Bill realised he might have looked confident on the outside but inside he was a blubbering mess. He needed fortification.

It was a tossup, which pub as he'd a choice of two across the road from each other. He weighed up. The Kings Head or the Royal Oak. Both famous since sixteen something, the Royal Oak was well known for far-flung loyalties of royalists named after the tree that sheltered King Charles after the battle of Worcester. Bill preferred the more down to earth clientele of the King's Head, which commemorated his beheading.

Both pubs boasted a welcome fire and simple inexpensive fare but Bill wasn't hungry and didn't intend to linger. He downed two ales, a whisky chaser, and left.

His cousin lived opposite where an attractive young lady lived, Alice Mary, the youngest of nine children. Bill used his cousin as an excuse to visit as he only had eyes for Alice.

Fearful with shyness, Bill had never approached Alice to talk with her. He knew he was being immature. He risked his neck and the lives of his crew on a daily basis, but was terrified of rejection by a pretty girl. The prospect of approaching her, only to be rebuffed and know that awful walk of shame back to his cousin was abhorrent to him. He'd experienced the feeling once at a NAAFI party, and it had unsettled him.

He had noticed Alice on his last leave but rather than approach her he'd talked inanely with his cousin while he watched her from across the street. She sat outside on her front steps. Her auburn hair had been held back in a colourful wide scarf, ends of which lay against her blouse. Slacks hugged the curves of her bottom, which did nothing to diminish the effect she had on his pulse. He felt a sudden kick of desire. It began with him wanting to remove her scarf and hold her close enough to run his fingers though her hair. Desire rushed through him.

He knew her sister Rose lived in the basement rooms and Alice lived upstairs with her mum on the top floor. Unbeknown to Bill, Alice had noticed him in his fine uniform. Her heart pounded enough to make her ears pop. Little did she realise what being aircrew represented to him. An unglamorous, inglorious and pitiless existence when night after night he pitched himself against horrendous barrages of flak. Wanting to see without being seen he cowered in freezing darkness, his eyes heavy lidded as he searched out any clue of what fate might have in store for him and his crew. He dare not relax, not even think of comforts like a soft clean bed. Death could come at them out of the darkness at any moment.

But Alice saw the glamour of a pilot and wondered, why doesn't he come and speak to me? What's the matter with him?

On that stormy afternoon perhaps his uniform became his anti-anxiety ensemble and gave him an uncharacteristic spurt of courage. In his favour there weren't too many pilots courting in the slums of Notting Hill.

Not ten minutes after Bill struck up a chat with his cousin and had lit his pipe, Alice appeared. She gave Bill a short, shy wave. He noticed how an angle of light reflected off her auburn hair and made it shine. Like following the rabbit in *Alice in Wonderland* Bill instinctively moved closer. His mouth formed 'Hello,' but nothing came out.

Over-whelmed by her curves he swallowed down the lump lodged in his throat.

'Hello, my name's Bill.' He hoped every sentence might not become an adventure in the woods.

Alice greeted Bill warmly. She shook her hair back off her temples and returned his smile with a sunnier one of her own. At last! And he seems better than some no-hoper poet without a pot to piss in.

'Hello, Bill, I'm Alice.' He stood open mouthed and grinning like a fool.

But she was transfixed. What might his lips feel like, she wondered?

Alice was not wearing a bra. Her firm nipples pressed against her blouse but something about her made Bill believe it was her innocence, rather than any attempt to show off.

She noticed his interest. She felt the flush at her throat spread and reached for her cardigan. 'It's a bit cooler today, Bill,' said Alice as she covered up.

It was downhill from there on in.

Alice was barely seventeen and had been nowhere. After school she'd become a trainee telephonist at the local exchange.

Her older sisters had 'been around,' their brains pulverised by motherhood, but Alice appeared greener than grass. Once they'd broken the ice there was no stopping them, Alice laughed at all the correct times and in no time they started to date.

Bill took Alice to the pictures and it wasn't long before, with the courage of desperation, they necked in the back row instead of watching Humphrey Bogart. He preferred the Gaumont as they'd thoughtfully removed some arms from between two seats in the back row. They enjoyed their closeness together.

'Oh, you are naughty, Bill. They're courting seats,' Alice looked coquettish.

First in the queue they quickly occupied their favourite seats at the rear of the one-shilling dress circle and left the majority to trudge into nine-penny stalls downstairs and miss out on the super seats. Alice felt great when she was with Bill.

She told her sister Kate, 'He treats me right and has his own money.'

'Perhaps then, you should hold on to him for a while.'

When they weren't necking they talked. Alice asked Bill about his life, his growing up and in particular his time in Canada. He told her how he'd discovered Sophie's plaster cast.

'Oh, Bill what an awful shock you must have had.'

Alice placed his hand directly onto her tummy, 'See, Bill, no plaster cast.' Her eyes sparkled pleased she made his loins twitch. But Alice had a problem. She liked Bill but had been warned by her sister Rose about going out with him. Had Alice been at all hazy about what men wanted, her older sister Rose was more than willing to put her right?

'Men are only after one thing, Alice. They'll tell a girl anything she wants to hear to get it.'

Another sister, Kate, was the eldest and married. She'd laughed like a drain when Alice confided in her. Kate had a rasping throaty voice from smoking too many cigarettes and drinking too much cheap gin.

'A wise woman makes her bloke wait,' Kate advised. 'If he's sincere and not wanting to shag and leave you, make him wait. It will strengthen his love, Alice. I should know. After a few mistakes I made my Stan wait.'

When Alice confided in Bill, none of this was what he wanted to hear. 'You're making no sense, Alice, it's not like that with us.'

This was one comparison between Alice and Sophie wasn't turning his way. He wanted to be with Alice as he had with Sophie before the awful pain of their break-up. But Alice was adamant. 'I'm not doing it until I'm married, Bill."

Alice wasn't taking any chances. She couldn't be sure, and all the time cautions of her older sisters rang in her ears. 'Have fun but keep your hand over your halfpenny,' they told her.

Sex to Alice was something she'd decided was sacred between husband and wife, which meant marriage and children. She saw no need to neither hurry nor flaunt it.

Bill backed off. He feared he might lose her. But it occurred to him he'd never sat and talked with a girl as if she was a gentle narcotic, and he liked it. He cherished Alice's company and felt comfortable and at ease with her. She was sexy and charming in an old-fashioned way and Bill liked that.

Alice neither smoked cigarettes nor drank alcohol, which Bill also found attractive.

That was an additional pull but demands of his work denied him access to Alice as in other jobs, life's uncertainty drained him further and being with her only increased his wanting of her.

Bill and Alice laughed and filled in their days this way, forgetful of the uncertainty of their lives. Days passed like hours when they were together. Bill felt as if he moved at the same pace as the sun in the sky. It was as if he could feel the earth turn.

Work ceased to be the challenge his energies needed to feed upon and instead took on the aspect of being a burden.

First time Kate had been warned by her mum about having sex with a man she was told, 'Remember, Kate, don't ruin your life for five minutes of pleasure.'

After she married Stan she wanted to know, 'Mum, how do we make it last for as long as five minutes?'

When a girl ceases to blush, she has lost the most powerful charm of her beauty...(Gregory I)

Chapter 29

DILEMMA

The nature of war is mayhem. Bill philosophised between ops as he and his crew worked it out. 'You know, greed and excess are only partly to blame,' said Joe.

'We bury ours, and they bury theirs. And everyone keeps doing it over and over again,' agreed Bazza.

'It's the mind numbing rapidity of us dropping our bombs while someone else racks up their cash registers,' said Bill. 'It's getting me down.'

'Aye, Bill, insanity of war has reached naught but its peak, by gum.'

'But not before more innocent lives than we choose to count are lost,' offered Mack.

On the wall was a blackboard with names of crews written in chalk. A corporal looked embarrassed as from his clipboard he began to rub out names of those who hadn't returned from the previous night's op.

'Pleased you're not rubbing our names out, Corp,' laughed Bill.

'Aye, too many lads are being sacrificed to the cause,' stated Ben. 'It's about time more value was put on the lives of our crews.'

'They should re-name Bomber Harris. He's Butcher Harris for sending us out too often,' said Joe.

Mortified the corporal broke off and left. Bill wondered if he'd completed his task or decided, because of animosity, to return later with less of an audience to complete his undertaking.

'In infantry, we'd be killed in some pointless skirmish,' sniped Frank. 'A few yards gained across a sea of senseless mud.'

'Attempting to take some hill of doubtful strategic importance you mean,' retaliated Joe. 'Troops and artillery shells pitched against one another.'

'And don't forget the poor bloody souls being told to defend it,' added Frank.

'Aye, but soldiers accept their lot more readily,' Ben said. 'We RAF fellows have naught but a more restless wee spirit.'

'Have you any idea what it would be like down there after we drop our bombs?' asked Sandy. He lowered his voice. 'It would be absolute bloody hell.'

'Same as when they drop their bombs on us,' said Joe indignantly.

'Like the devastation of Coventry 14th November 1940,' added Sandy.

'How come you remember the date? That was over three years ago,' said Joe.

'My Aunt Sally died in that attack. The bombers came through in waves, about 500 of them.'

Sandy became angry. 'Word is Churchill knew about that attack in advance, used it to impress the Yanks. Pity the Germans didn't bomb him! Pompous prick!'

Their group fell silent. Most knew of someone, a loved one, who had died in the bombing of Coventry.

'They say the firestorm could be seen lighting up the sky from over ten miles away.'

Sandy turned away, anxious his comrades-in-arms wouldn't detect his.

'There's no doubt, mates, we have unleashed the dogs of war,' announced Bazza.

Frank stared at the group. 'On the German side even more are dying than here,'

'About time,' said Joe. 'German High Command claimed Coventry was payback because we bombed Munich and Berlin.'

'Fuck the Germans,' Bill scowled. 'Sympathy! You'll find it in the dictionary between shit and syphilis.

They laughed then broke up as they'd things to do before their op. Bill especially wanted time with Alice. That the entire farce was being repeated over and over ad nauseam unsettled him.

Later in private, they petted but were not as intimate as Bill would have liked. Theirs was an old fashioned romance.

'When will this madness end, Alice?'

'I don't know, Bill. My brother George used to say how in times of fear and distress, decent human beings continue to do bad things to each other. Now Mummy is heartbroken and so am I. George has been reported missing in action in France.

'That's a terrible shame, Alice.'

'We'll never get over losing him. He was such a nice man, so soft and gentle. You would have liked him.'

Alice wiped a tear from her eye and Bill drew her close.

'I was at home with Mummy when the telegram came. The boy who delivered it cycled slowly up the street and looked at house numbers. I was really scared. I thought of you and then realised if it was you, the telegram would be delivered to Lily and Fred, not me, as I'm not next of kin. But rarely do telegrams contain good news, do they? It was as if the whole street hung in limbo. We waited to see which house he would stop at.

People appeared at front doors and watched from windows. We were all watching to see where he would stop. As he passed by each house, I sensed their relief. Thank God, not us, not today, not this time. I watched him and, oh, Bill, he stopped right outside our house. It was terrible. Mummy took the envelope and stared for ages before she opened it.

It read; 'Regret to inform you your son...'

Oh, whatever will become of us, Bill?'

Bill gently kissed Alice all over her face. Her tears tasted of salt and as he kissed, he licked the salt away.

'Much, much later, Alice, after fortunes have been made by a privileged few, and countless innocent people have been murdered in awful and excruciating circumstances, we'll be sold a new line—to kiss and make up. That's the nature of war.'

'You're so right. And you are clever. You see things others miss. Wars might be waged by nations but its human beings like us, who pay the ultimate price.'

Alice reached out for Bill's hand, 'Please, Bill, don't die,'

'I'm trying not to, Alice.'

Alice and Bill continued to pet but she was careful about how far she would let him go. She allowed him to loosen her bra and fondle her to her stocking tops. She felt his hands under her blouse on her back, smoothing and moving around to her breasts, and down to her belly. When he found her ribs, her hipbones, she turned her face and his lips found her mouth. He nudged at her legs with his hands and wanted her to open herself completely to him. But her loins were closed for business! And then they spent a lot of time gazing lovingly into each other's eyes.

At work Alice spoke guardedly with her friends but about little else other than Bill. Because of her pleasant nature most were pleased for her and hoped it might work out. Alice was a beautiful girl but her beauty also came from within and few failed to realise it. In effect she captivated them with her honesty, easy manner and lack of guile.

At Base, Bill was in a dilemma. He wanted to pour out his heart in letters to her, but thought it presumptuous. Deep down he was afraid to declare too much love for her in case it was spurned. When he was at home on leave without Alice, he felt desolate and totally infatuated. It seemed wherever he put himself he was in Lily's way. Exasperated she said,

'Stop moping around like a lovesick baboon and go to Alice.'

And so he did.

Alice had a little money put aside. She found a helpful professional photographer; Williams Pioneer Studios at Marble Arch. Their beautician covered up the worst of the sleep deprived grey circles under her eyes and fussed with the elegant sweep of her hair until she looked like a film star. The result was a postcard size for Bill to carry with him. When she gave it to him he was overjoyed.

'You look beautiful, Alice. I'll carry it always. I promise. Look, it fits snug in my tunic pocket.'

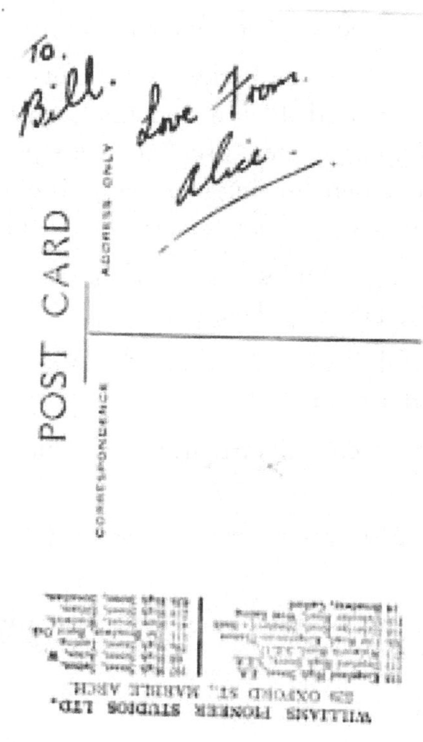
To.
Bill.
Love from
Alice.
POST CARD
ADDRESS ONLY
CORRESPONDENCE
WILLIAMS PIONEER STUDIOS LTD.,
229 OXFORD ST., MARBLE ARCH.

Bill's mouth had gone dry. He was overcome with emotion, but to him the photograph taken by a professional reflected only how Alice always looked. But Bill's attitudes to life were about to change. His will to survive fuelled by his sense of having been wronged, was intensified. He liked to think of himself as the cool headed, disciplined, rock-steady pilot but he began to realise, he wasn't. Instead he lusted over Alice's picture as if he was an adolescent. If a man was capable of making babies on his own, then Bill was a star performer.

He had always been ambitious, motivated by money and enormously driven but now, above everything, he wanted a simple thing; to survive and be with Alice. He'd had difficulty expressing his feelings in words but his hormones prompted his rebellious nature to surface in a myriad of ways. Butterflies in his stomach clamoured to escape.

At times it was as if his ambition to survive was in direct conflict with the hierarchy of the RAF. Not what they wanted him to do, but how they wanted him to do it.

So many planes had been lost and those green boxes signed out by pilots had become an expensive exercise. Word had got around, and according to the grape vine, when a bomber went down in flames it was a scramble by locals, who could get to the wreckage first and find Pandora's green box.

Because of this and purse cut backs, the contents of the Pandora box had been drastically reduced. It now contained approximately £30 in Dutch, French, or Belgian money. Even the Swiss penknives had been withdrawn. The powers were hurting financially. Those in the know, the big brass, wanted to protect their jobs and have more for themselves.

Bill's blatant breaking of rules was top of the list but he wasn't alone. The RAF was prone to breaking them too.

Bill was thoughtful. 'Lancasters were originally designed to carry a maximum bomb load of ten tons.'

'Aye, increased to twelve, then fourteen. Cookies weigh in at four. Then there's the extras.'

'Up to a dozen thirty pound incendiaries, with as many as twelve 1,000 pounders.'

Bill let out a sigh. 'An overfull bomb bay can carry upwards of twenty tons.'

'Aye, I don't know where they get summat like fourteen tons max from. I'm well in excess of that most nights.'

'To gain height I often fly nearly to Wales before setting course for Germany, Olly.'

Added to which we've got the usual mechanical malfunctions on overstretched Lancs, everything from faulty turrets to overheating engines.'

'Aye, gauges are unreliable, so called intelligence wildly inaccurate or just plain wrong.'

'As is the strength of winds forecast by Met Officers for navigators to make their calculations. Oxygen supplies are sadly depleted even before take-off.'

'Aye, if ground crew test them, testing itself often half-empties the cylinders.'

'Ben prefers to calculate on his hand-held Dalton compass rather than rely on sophisticated crap that could lead us further astray.'

Flying lower than instructed by his bosses had become a real problem for Bill but so far it had paid off with his life, and the lives of his crew. It also wasn't as cold although he barely noticed. But being lower there wasn't the same reliance on oxygen supplies and gunner's breach blocks might not ice-up so quickly. It gave them the option of being able to open fire if they had to. But fighters were no longer waiting on the outside as Tache had warned. Now they flew above or even among the flak, and the anti-aircraft fire was so intense Bill thought, hell, nobody could fly through that.

His Lanc was a slow lumbering vehicle and to make matters worse the Germans had faster more manoeuvrable fighter planes designed to bring him down.

And then there was that horrendous Schrage Musik, the weird German name given to their upward firing cannon. The most devastating loss-maker ever devised had taken countless numbers of Lancs but the RAF still continued to ignore its existence.

Despite a third of all German fighters being fitted with them, the RAF remained steadfast in their claim. 'It's a trick of the light. You're only seeing Scarecrow rockets, fireworks intended to appear as exploding aeroplanes,' announced Bomber Command.

'Right! So instead of trying to find ways of detecting the night fighter to protect crews and planes against it, top brass continue to take the easy way out, by fabricating stories and telling lies,' raged Bill.

'Aye, smoke and mirrors don't take Lancs and entire crews. Those lying pricks at Bomber Command have been circling the drain they created for themselves ever since they were put in charge!' thundered Olly. 'It's naught but a fuckin' disgrace.'

Bill had learned during training how when a plane flew closer to the ground it made a more difficult target for the enemy to hit, provided a Schrage Musik wasn't already underneath it. If it was then it was difficult if not impossible to protect against. He couldn't always fly so close to the ground as to prevent a fighter from flying underneath him. If op instruction was to fly in at a bombing ceiling of 20,000 to 22,000 feet, it wasn't unusual for him to fly lower, but there were restraints. Not too fly so low as to affect their camera flash.

Lancs were supposed to group together for safety in numbers. But after they'd dropped their loads of bombs, Bill often returned home alone and perilously close to ground. Sometimes too low for a corkscrew manoeuvre or bale out. While he changed things about to suit himself Bomber Command were paranoid about everything in tidy neat boxes.

'Aye, it's like they're organising a bloody parade rather than bombing Germany.'

'You're not wrong, Olly.'

Bill's plane often crossed the target at minimum height for bombs to detonate and pictures to develop without hurting him or his crew. He liked below 8,000 feet. Often he disobeyed instructions by breaking formation so bombs dropped from planes above didn't fall on top of him.

Returning home at night he often flew at less than 1,000 feet. He acknowledged this was in breach of RAF instructions, particularly over water where he'd no horizon to follow. Even in daylight any horizon was every imaginable variation of a dreary colour grey. Nearer home he needed more height. There were other hazards, which didn't differentiate between friendly or enemy planes. Dozens of barrage balloons floated in the sky above British cities, moored by thick steel cables. In inclement weather the unwary risked shearing off a wing.

But so far these strategies had worked for Bill and his crew. He was alive. They were all alive. He was holding Alice's hand now, able to gaze into her beautiful big brown eyes, which shone back at him. She was nibbling her lower lip, again. A pout. Alice moaned and reached wide her arms, her exposed throat, she became like a cat at full stretch. Then she changed tack and fiddled with her hair. Bill wanted to be with Alice and she wanted to be with him, which meant he'd grown to dislike mostly everything about the RAF.

Alice being straight-laced, prudish and a virgin didn't help matters. Bill wanted her to be as accommodating, as Sophie had been. Alice began to waiver. It wasn't long before she wanted to be intimate with Bill. But compliant as Alice wanted to be, she remained adamant. 'Before we do it, Bill, having sex, I mean, we have to be married.'

Bill was pleased Alice was not promiscuous. He didn't want to share her favours with any other man but he also didn't want to wait. When he cast his mind back to Sophie he realised although they'd had a proper relationship, he could never have trusted her, as he could Alice, who by comparison was chaste.

Bill was twenty-one years old but war had aged him. It made lives uncertain. What if they did marry? What if Alice became pregnant? A child born out of wedlock was a bastard. To be illegitimate meant not nice. A person born of ill repute. And what if Bill perished on an op? What then? They both wanted to be lovers. Bill had found the woman he wanted to spend the rest of his life with but should they take the plunge? Would it be fair to Alice if Bill were killed?

They weren't the only ones wrestling with these issues. All over Europe young couples faced the same dilemma.

Alice getting on well with Lily and Fred helped tremendously. She'd become a regular visitor at Barlby Road. 'You're as welcome here, as you are at home, Alice,' Lily said. And she meant it.

When Bill was on leave Alice stayed over but they never shared the same room or bed. Alice occupied Bill's room, the little room, on her own. She lay there throughout the night half hoping he might break her cardinal rule and visit her. She longed to hear his hand on the door handle and see his shadow in the doorway before he tiptoed across the cold linoleum to clamber into the featherbed beside her. Would his feet be cold? She imagined his strong young arms holding her pressed against the dishevelled quilt bedspread. His keen impatience as their kissing became urgent, the desperate way he'd remove his clothes to be naked next to her. Should she assist him or might she appear too brazen? She began to visualise how she might arch her body allowing him to work her nightie up around her throat. Should she take his penis in her hand as Kate suggested, or let him find his own way?

Rose laughed and blew cigarette smoke towards the ceiling when she'd said, 'if there's hair around it, he'll soon find it, love.'

Alice continued to wonder and fantasise. Might Bill lie on top of her and she widen her legs, or would he turn her over and spread the cheeks of her arse. Kate had told her, 'taking a man from behind is best, love. It's more sensitive for the woman.'

Secretly Alice hoped Bill might have gained sufficient experience as to lead her but irrespective of what position he took, she smiled inwardly at thoughts of how he would hold her close until she lost her virginity. She wondered what Bill's penis might feel like as she'd only ever explored with her own fingers.

But she would have to wait to find out as Bill never granted Alice a nocturnal visit. He trod the corridor a few times from the scullery but only to the toilet. He tried not to disturb anyone by dodging boards that creaked. And when he passed his bedroom door he paused, looked at the door handle in the gloom, and wished he had the nerve to open it, risk her displeasure and venture in.

Little did he realise the love of his life, on the other side of that same door had heard his gentle tread. Alice had almost stopped breathing with anticipation. She held his sheets up tight under her nose to absorb his scent. She fingered herself lightly under the covers as she sniffed him, and she fantasised about his strong young body entwined within hers. How she would hold him tight, kiss him all night and never let him go until she'd drained him dry. Alice sobbed herself to sleep with the passion of wanting Bill.

But Bill kept his word and stayed away. He remained at his lone station, the armchair in their scullery. He sat beside the warm hob, and dreamed of the day he would consummate his love with Alice. He would do it legitimately without the need to wear a condom and he planned to do it to her in every position his mind fantasised. *Pick a hole, any hole!*

Lily and Fred embraced Alice as if she was the daughter they'd never had. And the daughter-in-law they longed for. They both wanted to be proud grandparents.

Days in between Bill meeting up with Alice dragged terribly but most days in London were miserable. But then for Bill, when Alice was at his side, nothing else mattered. Shortages, long queues for food, unfair and inefficient systems of Ministry of Food ration coupons, the black market, not to mention the constant fear of air raids, mutilation and death all paled into insignificance for Bill when he was with Alice.

The Women's Land Army was in the news. They wanted girls aged from sixteen. Green jerseyed, dungaree girls worked fifty-hour weeks doing jobs vacated by men who had been called up to fight.

Alice wanted to join them. Bill became distraught.

'Why? Do you like the drab look of their baggy overalls?'

'No I don't. But over 80,000 strong can't be wrong, Bill. And surely it's better than being assigned to a munitions factory.'

Fred peered over the top of his newspaper. 'God help us. You don't want to be assigned there, love. Too dangerous, too many accidents by far.'

'Dad's right, Alice, don't go there,' pleaded Bill.

The Women's Land Army was paid by individual employers, not much, about £1 a week, less if they lived in. Cheap untrained labour had to learn on the job.

And then came the clincher from Fred. 'You realise they could send you anywhere, Alice.'

'What's the point if we can't be together?' scowled Bill.

Lily agreed with Bill. 'The money's nothing to write home about, Alice.'

Fred struggled on his railway pay, which didn't help matters. Dazed with misery and fatigue Lily charred for pin money. Between them they only scraped by.

The black market thrived, paid for by sleight of hand and slippery practices but without financial assistance from Bill, it's doubtful any of them would have eaten an egg or put milk or sugar in their tea.

When Bill was at Base he lived well. At home on leave their rations were meagre. They ate whatever they could get that was plentiful and cheap. Lily became more adept than ever at baking and showed Alice. Together they tried new things. Parkin, a filling ginger cake made with oatmeal and treacle, combined a chewy texture of oatmeal with sweetness of treacle and a spicy bite of ginger. Their problem was finding Snowflake self-raising flour. Rarely did they add an egg or butter. Canned salmon became a luxury.

Before Bill met Alice on ops he'd psychologically written himself off. Now he had Alice to come home to, he could no longer do that, which posed a problem.

His mind, even while flying his Lanc, sometimes wandered, filled only with thoughts of Alice. He couldn't bring himself to discount his life the way he had before he met her. He wanted more than ever to survive his dreadful ops, the war to be over and for him to be at home with Alice.

Those who could afford to get out of London did. They moved into quieter country areas away from air raids, bombs and mayhem.

Bill summed up. 'Winston Churchill might be popular with some but he isn't popular with everyone. The war mongering bastard said if the RAF only lost one hundred Lancasters in a raid, it would be worth it. Arsehole!'

'Are you sure, Bill? Goodness that's 700 brave men.'

'I'm absolutely sure. It's on record. He said losses of ten percent were acceptable.'

Fred disliked Churchill with a passion, as did Lily. They saw a bigot, a man who had been born, not with a silver spoon in his mouth, but a big golden one.

Lily's head was down as she knitted. 'With his wealthy family connections he's done well, but then you'd expect that.'

'Manoeuvred his way into powerful top jobs at the expense of thousands upon thousands of innocent peoples' lives you mean,' Fred continued, 'I don't understand what people see in him. He stuffed up Gallipoli, can't even hold his marriage together, Girl. They'd have been better off putting Len Hutton in charge, than him.'

Lily paused with her knitting and raised an eyebrow at Fred's potty language. Alice ignored it. She was used to colourful language as Bill had a serious mouth on him.

'She, Mrs Churchill, spends most of her time off in America where it's safe,' Lily continued. 'But she'd want rocks in her head preferring to be with him. Big fat slob he is. And always ponging of those big fat cigars.'

'Not if she can find a good looking wealthy American to shack up with, Girl,' laughed Fred.

'Yes, just think. Fifth Avenue and Tiffany's.'

'Churchill's not the only warmongering inbred who has repeatedly stuffed things up,' said Bill. 'We English celebrate our defeats and failed invasions with more passion than our successes. How we love our legends; like the battle of Balaclava in the Crimean War and the ill founded charge of The Light Brigade. Could only be described as a suicidal charge under the irresponsible command of Lord Cardigan.'

'Who was Lord Cardigan? What's he about?' asked Fred.

'He was the drunken slob of an aristocrat in charge. So pissed, so foolish, so up himself he sent the Brigade charging down onto the wrong guns. Bravely executed their charge was a complete fuck up. It convinced the Russians how truly mad we English were, but apart from that, it was a typical balls up. The brigade received conflicting orders and charged off—in the wrong direction—and at the wrong guns! Come to think of it, sounds like something my RAF bosses would conjure up.'

'What happened, Son?'

'It was intended they chase down on the Turks to stop them carrying off valuable British cannon, a simple and safe task. The cannon were already partly dismantled making firing them impossible. Most likely the Turks would have scattered in the path of the Brigade. Instead, Lord Cardigan supposedly misheard his orders, became confused poor bugger and sent the Brigade down on Russian cannons, which were active.'

'What became of Lord Cardigan?' asked Fred.

'Richard told me it's doubtful he led the charge. There was talk about him being on his yacht for a champagne dinner in Balaclava Bay. He returned to England and blamed everyone else for the cock up and was promoted.'

Everyone shook their heads in dismay.

'A case of let's not think too much about what we're doing for fear the noise of our thinking might scare the horses, Son.'

As George Orwell put it, *we didn't know how bad it was until they told us.*

Lily and Fred feared for the welfare of kids in London, and thanked God they didn't have any young ones to worry about. Children were being taken forcibly from their families for their own good. Adorned with large nametags and addressed like parcels as the refugees they were about to become, they were sent to foster homes for protection from ceaseless bombing and carnage.

Some children fared well, others all right, but most not good. Of those who fell into the wrong hands, and there were many, some would be psychologically scarred for life. Many were never reunited again with their own families who had loved them. Others would become adult without knowing who their real families were.

Alice became distressed. 'Promise me, Bill. If I don't survive this awful war and you do, you'll never let our baby when we have one go to a foster home.'

Bill promised.

And then there were suspicions. They were everywhere. Good citizens, even those who had lived in Britain for decades, if they had a hint of foreign origins were rounded up like cattle and interned. Placed in detention camps they were treated worse than POWs. Their properties were frequently seized to assist the war effort.

'Even Daschund dogs are being viewed with suspicion,' snorted Bill. 'How bloody stupid is that?'

Lily made a smile that looked as though she had toothache. 'And so they should be. They're half a dog high, dog and a half long and look stupid.'

'We get a laugh out of them for humping our legs, Girl. But if they're rich, even foreign, they can get anything they want at a price.'

'Not every politician is rich, but they do seem to vote themselves privileges and access to things we don't get.'

'Comfortable shelters that don't fill up with water when it rains, for one,' reminded Alice.

'And the best of food, alcohol, and even big fat cigars to smoke,' chimed Bill.

'For them this isn't a war. It's more of a bloody inconvenience played out for their benefit by the likes of us. They sit in a comfortable armchair by a wireless, waiting as a servant runs their fresh bath.'

'True. If you're poor, you have little choice but to join endless queues at every shop, clutching your Ministry of Food ration book and going without in aid of the war effort,' moaned Lily.

Week after week, month in month out, the hierarchy recited their prepared speeches. They delivered the same killer pitch over and over to prolong the momentum. It was as if their principal concern was people en masse might tire of this madness, lives of drudgery without small luxuries. They might prefer any alternative as long as it was untainted by anything that smacked of being foreign.

Their speeches intensified with the carnage. 'To fight in the streets, and on the beaches,' rather than be subjected to German ways. The tune was called perfectly, the piper paid generously, and in most cases those in charge fared well.

'None of the politicians has any intention of being engaged in hand to hand combat anywhere. But they know they will never have to,' raged Bill. 'Not when they can send others to do their bidding.'

Chapter 30

WEDDING 21 MAY 1944

When he was home, every morning was as if the sky was scrubbed clean. Bill found a neat pile of clothing laid out for him by Alice. Always a fresh folded shirt, socks and even a handkerchief had been pressed. Alice insisted she prepare Bill's meals and Lily was content for her to do that. If Bill was late home, which was often, Alice served him his dinner. Then instead of leaving him to go and sit with Lily and Fred, she sat with Bill and watched him eat. Bill had never felt more superior and wanted.

He mentioned this to Fred, about how girls might behave after marriage.

'Normally,' said Fred, 'I'd say take a bloody good look at their mother first. But in this case, looking at Fanny's not much to go on. Perhaps have a passing glance at how she acts towards her father. In Alice's case she hasn't got one, so you bomb out there too. I suppose you've struck it lucky, Son.'

Bill thought long and hard. The best part of getting home after work was the look on Alice's face. When she saw him and knew he was safe she was beside herself with joy. She knew for a while at least, no telegram was on its way, which made her smile for days.

Again Bill took solace from his favourite childhood fable *Alice in Wonderland* and followed the rabbit when he asked Alice to marry him. 'I love you so much, Alice.'

'I love you too, Bill.' And she did with a fierceness that hurt. She said, 'Yes,' and Bill's throat clogged with delight.

Their decision was made. They chose to ignore the madness and become man and wife. Neither had been married before and Bill had nothing in his past experience to go by but it began to dawn on him that when he married Alice he would be nicely fixed contentment wise.

Talking of material things, Alice being old school there were other aspects. As soon as they'd announced a date Alice set about studying Bill's comforts. Her goal was a good marriage.

'Aye, that attitude's beginning to go out of fashion with younger women, but not with Alice. There'll be moves afoot for sure. Women to earn same as men and have equal rights but that'll make no difference to her.'

Bill agreed.

Lily and Fred were delighted with the match, as was Alice's mum, Fanny.

But where should they marry? Bill had grown up as Church of England.

'That's the denomination we've always been, Son, because we're really no denomination at all,' laughed Fred.

'We are Christians,' Lily added, 'but only for weddings, funerals and christenings. We don't go along on Sundays to sing hymns to a God we don't believe in.'

Being Church of England was an acceptable response to any question asked about someone's religion but few took it seriously as other faiths. Alice was not religious nor were any members of her family, which simplified things. Bill had never sought spiritual assistance. he'd thought that necessary for others, not him. He was convinced, as were Lily and Fred; the good news was there was no hell, the bad news—there wasn't any heaven, either. Their family were atheists and proud of it. In that sense they were irreproachable.

Bill became serious. 'God loves violence, otherwise why would there be so much of it.'

'If you ask me He should be ashamed and cast himself out for eternity for the shocking balls up,' countered Fred.

'Ideas of an afterlife, in point of sheer logistics, do seem pretty far-fetched,' said Alice.

They agreed not to be conned by 2000 years old de-ranged fairy stories.

'If they're right and there is a heaven,' Fred said. 'It must be getting seriously crowded by now.'

'I'd prefer not to go,' said Lily, 'I don't like many of the people who are already there.'

After a short engagement, without a ring as money was tight and times were tough, Bill married Alice in a Registry Office in Royal Borough of Kensington on 21 May 1944.

They all frocked up. Fred in his best and only suit. His neck protruded trough a collar several sizes too small for him. He looked the epitome of a solid citizen. His watery blue eyes and fleshless cheeks were accentuated by his nervous tic. His hair was trimmed and combed in a tidy sweep across his forehead. Lily wore her heavy coat over her best dress and a wide brimmed trilby hat.

Olly was best man. 'Aye, Bill, a lot of young men including me were in war for summat to do, uniforms and stripes attract girls by gum, and you got yours for sure.'

Alice wore a pale blue coat to below the knee and a hat that suited the occasion while Bill wore his dress uniform.

Their wedding was simple, straightforward and uncomplicated. Family and friends chipped in afterwards for a feast at 8 Barlby Road. Olly toasted the bride and Uncle Charlie toasted the groom while they arranged themselves around Lily's best table in their rarely used front room, adorned with its pressed white linen tablecloth, doilies under the vases and polished silverware from their bottom draw.

'If only for the moment, we feel rulers of our own special empire, Son,' said Lily.

'Certainly an improvement on last year's shambles,' laughed Fred.

Fanny agreed with the testy bark of an old biddy and they all sang along with Bill and Alice's favourite September Song while Lily and Fred wept to see the result of their handiwork as matchmakers,

On their wedding night Alice and Bill couldn't keep their hands off each other. Under cover of their darkened room, accentuated by heavy blackout drapes, even the stars struggled to get in. Alice lay naked in Bill's arms. All her fantasies of being an alluring female, a goddess, vanished in a flash. Bill was on her in a race to the finish. They devoured each other in a floodtide of passion. Afterwards, Alice hoped she might have conceived.

There were a few brief days of harmony before the next wave of aircraft droned overhead and sirens wailed. They heralded a stark reminder they were back into the realities of war.

Bill prepared mentally for his op and adjusted his expression to a careful blankness.

At 0600 hours he lay awake in bed between clean sheets with Alice and pondered. Would he wake up with Alice in two days time?

He attended to his haemorrhoids, washed, shaved and looked in the mirror. The man who stared back surprised him. He half expected to see the image of a forty-year old. Instead, he saw this person. Would he gaze in the same mirror at his face in two days time? Thoughts that he might not saddened him. Not that he liked his own face but he instinctively reached out and touched the mirror. As he did so he realised there were tears in his eyes. Bill composed himself and dressed.

On the way out he paused and turned to wave to Alice. Whenever he left she always stood at the same upstairs front window. He waved back, turned and felt embarrassed at the fuss. He was unsure why but realised in that moment he saw a sparrow in a tree. There was nothing special about the bird or the tree. But again, he thought might he be standing there in two days time?

Bill knew that night he would fly into some hell on earth somewhere over Germany. The prospect of him being shot down was high but no one spoke about any of that.

Another pilot had told Bill back in the Mess. 'If you see an orange flame in the heart of the explosion, it's too bloody close, Bill.'

As take-off hour approached, twilight seemed to hold on forever as day receded into night. Bill had never thought much about twilight as a child but now acknowledged how much he despised dusk. Perhaps it was the insecurity, a period of danger when the Luftwaffe might spot his Lanc before total darkness enveloped him. Whatever the reason, he'd grown to dislike twilight with a passion. * 6th Footnote

Take-off was set for 2000 hours. Further delays meant the wrong side of 2100 hours before they were in their queue. 'If they don't get a move on, it will be getting light over the target,' grumbled Bill.

After take-off, another irregular load meant having to fly almost to Wales to gain height before Ben said, 'Swing the wee plane back around 180 degrees, Bill.'

After crossing the English coast at less than 4,000 feet he began to wonder if he would gain altitude at all. Might the armourers on this occasion be his undoing? Had Joe signed paperwork for a consignment of bombs devoid of weights? Had ground crew themselves been unsure? Maybe after weights had been finalised some wise arse found room for more? He thought briefly of the consequences of dropping his bombs in the sea but ruled out the idea because of the camera flash and his height. At less than 4,000 feet if the cookie exploded as it hit the water, the explosion could take him and his plane. He had no option but to gain more height.

On his first op he'd felt sorry for those crew who were not able to see what was going on around them. It dawned on him now perhaps they were the lucky ones. Sometimes he wished he could close his eyes and keep them closed as what he saw traumatised him to the core.

Slowly the big plane climbed, but not without effort.

Lower than intended over their target, flak was intense but shells burst above him.

The thought of the Hun not finding his measure amused him. He'd never noticed before but the explosions seemed to come in bursts of four from each anti-aircraft battery. They made an interesting group pattern. When he saw a shell explode he began to count on another three to come.

It became a dangerous game. If he saw two, he'd guess there might be another two to come. In the past Bill had dismissed flak as simply being there, something he could do nothing about. Now he began to wonder at its predictability.

A bonus from heavier flak might mean fewer enemy fighters. What if Tache had been right? All the time anti-aircraft shells kept coming German night fighters might stay away? He fidgeted. What if a German night fighter was already positioned beneath him? What if it had already identified his silhouette and was armed with their feared Schrage Musik? For all he knew it was already taking aim at his fuel tanks. Bill drew deep breaths and began to feel sick. *I'm thinking too much, that's the trouble. I'm putting too much bloody thought into it.*

Bill wondered if his plane erupted in flames whether he might know much about it. He reasoned there would be no memory of the pain, so perhaps not. He tried to push black thoughts of negativity from his mind. What if he was underneath another Lanc when it dropped its bombs?

Fuck! Never was he so pleased to drop his own bombs and get away. With bombs gone the plane was lighter and responded better to his controls but he realised, prior to that with his heavier load, any successful corkscrew manoeuvre might have been beyond their capabilities.

Bill saw four Lancs go down over their target. There were others lost he never saw. On his return trip he watched what looked like a bright pinpoint of light in the darkness ahead of him. It was difficult to identify. Too bright for an exhaust and why only one when a Lanc has four?

The light continued to fascinate him. It stayed constant for a while as it moved ahead of him through darkness.

Fuck! It must be moving.

Then the glow began to grow in size, and as it did, it became brighter. Then it changed shape. Suddenly it erupted into a great ball of white fire. Flames became visible but changed form as it fell. Bill watched it go down much as a child might observe a dying Roman candle firework.

Bill realised too late he'd watched a Lanc in trouble. Hit and in flames it had flown on until the inevitable. Another Lanc and crew shot down. This one had been in front of his eyes. That could easily have been him. He wondered if a chinless wonder in intelligence might berate him later, if for him there was a later, for not getting its number.

At de-briefing there were more tears than usual. Those who didn't cry were doing a lot of loud nose blowing into their handkerchiefs. He found himself pretending not to notice. Then realised he felt nothing. For him no tears, no blowing hard into his handkerchief. There was only relief to be back. He looked about and noted absence of many familiar faces. The pilot who had spoken to him before take-off simply wasn't there.

So few crews had returned home.

Men sobbed. They appeared to be younger; perhaps it had been their first action.

He felt sorry for them but was aware there was nothing he could do. Intelligence officers appeared to pay them no heed, but made copious notes. *Heartless, bastards!*

After de-briefing Bill and his crew devoured their celebratory meals and Bill felt embarrassed and ashamed. So few were there to share the moment. Offered second helpings he was suddenly overfull. Saddened, he departed for home.

At 0600 hours Bill awoke. Pay day. Rent day tomorrow.

He lay in bed between clean sheets next to Alice and mused over his thoughts from forty-eight hours ago. He'd wondered if he might awake in the same bed with Alice in two days time? Remarkably he had.

Bill washed, shaved, attended to his haemorrhoids and stared in the same mirror.

Tonight was another op; would he look into the same mirror, at the same face in another two-day's time? Thoughts that he might not depressed him but not as much as not waking up next to Alice. He'd grown to like her face more than his own. He instinctively reached out and touched the mirror and as he did so he realised Alice was standing behind him. Tears streamed down her face. He spoke to her reflection in the mirror.

'What's up, Alice, why so sad?' But he already knew the answer to his question.

'It's nothing, Bill,' said Alice, as she composed herself in the glass. She ducked, as if shy of her own reflection but they both knew why Alice had shed tears.

Bill helped compose Alice with a hug, which only made it worse. Then he dressed. Alice tried to arrange her face into a forced expression of lighter mood but it was an effort.

'It's not as if we have a choice, Bill.'

Bill agreed. On the way out he paused and turned to wave to her. She was at the same window. Then he looked for the sparrow in the tree. The tree was there but no sparrow.

Oh, fuck! He panicked, again. Found himself searching the sky for a bird, any bird.

Thank Christ! He found one circling above. But now it looked like a flying dagger. *Oh, fuck!* Was it a bad omen? The Fokker Wolf was known as the Butcher-bird among RAF crews, and with good reason due to its manoeuvrability and speed. Bill was desperate to clear his head of negative thoughts. He found himself hoping the absence of the bird was not another omen but if it was, he accepted there was nothing he could do about it.

I keep this up next I'll have voices in my bloody head. He quickened his step and as he did, Alice's handprint faded from the glass.

He tried to push thoughts and horror images of tonight's op from his mind. He felt like a humble farmer might, if yoked to a wealthy man's plough.

If we cannot find a God that suits us, then we should invent one... (Voltaire)

Chapter 31

FAMILY TIES

Bill continued to grow apart from his dad with cavalier indifference. He was angry Fred had missed out on too many good opportunities to better himself. Annoyance of it gave him toothache.

'You should have profited more from your non-combatant role during the war,' yelled Bill. 'So many prosper. Why not you? For them business isn't good, it's spectacular.'

True the black market had raged, but not for Fred. It was a major disappointment for Bill as he lay beside Alice in bed and shuffled through his worries about money.

'Trouble is, Alice. We never seem to have enough.'

'Money isn't everything, Bill.'

'I know. But if money is the root of all-evil as we have been taught, then why do churches and England's elite have so much of it?'

Fred's long-standing fondness for strangers, his tendency to befriend any new face, further irritated Bill. It hadn't worked for him. Every time he became attached to anyone, even dog Scruffy it was as if some greater power decided to take them away, by killing them off. Fred espoused the virtues of his new friends while Bill tried hard not to become too attached to anyone except Alice.

'It's not hard,' raved Bill to Fred. 'You don't have to be brilliant. You don't even have to be clever. Just have a bloody pulse and be motivated to turn up. All you have to do, you silly old fool, is be there!'

While Fred gazed into the distance with a look of resignation that summarised his life, Bill knew there was always someone desperate for whatever could be provided. Anyone in legitimate business found themselves in a position to profit from the war effort in ways they'd never thought possible. Anyone not actively engaged in combatant duties like Fred, should according to his son, have become rich during war, if only he'd suitably applied himself.

'Profiteering is only a dirty word at the lower end of the scale,' raged Bill.

In a city in need of everything Bill saw only opportunities, as everything was in short supply. A small allotment for growing vegetables was a gold mine. It needed to be watched over and guarded against opportunists because as soon as a glimmer of anything green raised its head, it would be taken. Nothing much was available anywhere.

Bill was hopelessly obsessed with money and appearances to the point of being allergic to every simple kind of fun while his parents were hideously conformist.

News of U-boats sending merchant ships to the bottom of the sea was daily news.

Briefly Bill wondered about Jack. As a stoker might he have gone down with a ship?

Bill shivered. He'd never liked the man, but he didn't wish that on him.

Second hand anything and everything resold at a premium. Toys weren't available even at Christmas. Shops were mostly empty. Millions of tons of goods found their way to the bottom of the Atlantic Ocean over a six-year conflict, attributable to a brutal, terrifying cat-and-mouse struggle between ships, aircraft and German U-boats that ranged over thousands of square miles.

Pity we can't bomb a few more of those German submarines, thought Bill.

The Germans knew if they cut supply lines they could win the war. Steel, food, clothing everything for Britain's survival was going down daily by thousands of tons in convoys from America.

Kids wanted toys at Christmas and there weren't any. Toy soldiers were of unpainted lead from the same mould. They were identical. Where the lead came from or who made them no one cared. If children sucked them, no one fussed. Fred had made some himself. He'd melted lead in a mould on the hob and cooled them on their kitchen table. Then he'd given them away to children of friends.

Bill berated him. 'Why not make more and—sell them!'

Lily and Fanny 'charred' for additional pennies and that tore at Bill's guts. To him it was about being in that class conscious society again, how some were born to serve, to do menial jobs for the elite while to Bill, elite meant anyone on the next rung up. The amazing thing to Bill was how the lower working classes seemed so happy and content but it rankled.

Bill knew he was out of step, but for the first time in his life he had someone beside him of like mind. Alice was unhappy with Notting Hill, hated life there with a passion, and longed for better. Like Bill, she wanted out.

Not that Lily and Fred were deliriously happy. They would have been over the moon if blessed with a windfall like a big win on the football pools. But otherwise they were content with their lot. Bill wasn't. He struggled with his inner ambitions.

Part of Lily and Fred's contentment stemmed from surrounding themselves with people who didn't think in any more depth about life than they did. And that also riled Bill.

Alice and Bill schemed and plotted to leave the drudgery.

After they married Alice had moved into Bill's little room at 8 Barlby Road. And when Bill came home on leave they spent time together.

Soon they became unhappy being apart. Bill realised Alice was the sort of wife who liked looking after him to the exclusion of everything else. He was lucky. When Bill went back to his aerodrome, Alice wanted to be with him.

'Here are your sandwiches, Bill.' Alice put a package of something wrapped in greaseproof paper in his hand. He kissed her and descended the single flight of stairs out into Barlby Road. As he walked towards the bus stop he glanced back. Alice stood at the front room window and waved as always.

Bill had found peace of mind. He forced himself not to look at the trees or sparrows. Only Alice. He never dwelled on thoughts of Sophie again.

When he went to work, which was how he described his job as a bomber pilot, he never considered for a single moment Alice's movements or interests, outside of them being together. This gave Bill great contentment and stability beyond any of life's needs. But they wanted more time with each other. He became totally endowed with her physical and mental charms, her innocence of mind. Together they were as one.

Apart from any sexual passion, Alice was staunchly proud of her young husband's achievements in how he'd become a pilot with such limited schooling.

It wasn't long before Alice, to ensure they could be together through every occasion, took lodgings close to wherever aerodrome Bill was based. Their first was a room in a farmhouse.

Now Bill had only one vision; a warm, smooth-sheeted bed in a darkened room and Alice lying naked and expectant within it. After Bill returned from work he found Alice and she became molten in his bed. He explored her body gently, but as greedily as a desert wanderer savaging a found bunch of grapes. They lay together as Bill pumped his passion urgently and repeatedly into his young wife, until after a while it became mechanical.

Alice wanted a baby but rarely showed any sign of climax. She lay there night after night for Bill's benefit. Sometimes if there was sufficient light in the room Bill would cast a sideways look at his beautiful young bride. Her hair lay over the pillow but her eyes were wide open, as if transfixed on a single point at the ceiling rose.

The universal comment, 'lay back and think of England,' came to mind.

Then one night, while making love Bill stopped. She panicked. 'What's the matter, Bill?'

'Nothing, Alice, but we're going to try something different.' Alice climaxed with a man for the first time in her life. Later while they ate their meal together she laughed. 'Guess what, Bill? Now I know what my sister Kate meant,' Alice said coyly.

'What's that?'

'About trying to make it last at least five minutes,' giggled Alice.

It wasn't long before Bill was called in to see an officer-in-charge and told to send Alice away.

'It's against regulations for you to co-habit while you're on active duty.'

Bill was insubordinate. 'If you don't want her here, Sir, I suggest you go tell her yourself.'

Alice was never approached nor asked to move away. She continued to follow Bill around the country to be near him. But lodgings in rural areas brought a high premium. Country folk weren't daft. When Alice followed Bill from aerodrome to aerodrome. 'Nothing is given away for naught,' they said.

This was the beginning of Bill's lifetime of serious disagreement with officialdom and bureaucracy. For him they'd gone too far. His contempt for anyone who led an unworthy or privileged existence, ensconced above what he considered their fair pay rate should be, was more than a little obvious. He was never backward to share his thoughts about inbreeds of royalty and how many aristocratic families were gutless bureaucratic swine.

'That a nobleman had carried out some deed for his King, something of such value he curried favour and had bestowed upon him lands, wealth and title was one thing,' said Bill. 'But that his inbreeds live off his endowment for centuries to come is quite another.'

Bill painted a vivid picture of how those at the top should serve His Majesty in a more honourable way, or at least, be more considerate for the scared teenagers who did.

'Them and their power brokers are a blot on the landscape. Their underlings, the civil servants, are afraid to do the right thing and act without even mandatory civility,' ranted Bill.

Alice was in full agreement.

'Coming from our backgrounds, Alice, I've never quite been able to grasp why people take the royal family seriously.

'I like the royals, Bill. I think they're glamorous and exciting but outside of that a couple doesn't need a social life when they have each other.'

Further confirmation I have found my soul mate, thought Bill.

Time goes by a lot slower when you miss the one you love... (Anon.)

CHAPTER 32

DOODLEBUGS

After the Blitz came the German V1 Rocket a.k.a. the Doodlebug. Needless to say most Londoners were terrified.

'Designed by some clever arsehole to fall wherever it runs out of fuel,' raved Bill.

It caused multiple mood swings. Most were gripped by cold dread. * 10th Footnote

'Our life expectancy can depend on as little as an extra cupful of kerosene, Girl.' 'Keeping your head down and your powder dry is little help against them.'

'You can tell the difference in sounds between doodlebugs and other planes,' said Lily. 'Bombers are louder.'

Fred took Lily to task. 'Of course they're louder. They've more motors.'

'Apart from that bombers sound different, less constant in pitch. Dreaded doodlebugs make a sound more like a truck with a broken silencer,' laughed Lily.

Bill reminded his parents. 'It's powered by a pulse-jet but they're all life-threatening noises, you two.'

But Lily was right. Of course the kerosene powered engine sounded different.

Irrespective, they were unnerved by the doodlebugs. Unpiloted, they were random death machines. If bombers dropped their loads a few streets away they'd try to laugh it off.

'That German bugger's missed us again. He's way off from East End docks, Girl.'

But when they heard a doodlebug approach it was hard to ignore, especially when it droned closer like the buzz of a thousand bees. Their pulses quickened. Overhead it was the loudest. They closed their eyes in anticipation of unspeakable horrors to come. Then as it passed by, they relaxed a little. Knowing if it fell, the impetus should carry it onwards and a little further away from them. Then, as the sound faded, they sighed with relief.

'What a bastard of a way to live, Girl.'

Lily was in tears.

If when a doodlebug approached they heard the engine suddenly cut out they panicked. That was the point at which it began the last stage of its murderous freefall. When it hit the ground it exploded. Huddled under their Morrison shelter they clutched each other in the gloom. Fred's nerves were at breaking point.

'I don't mind death, Girl, as much as the fear of dying. I know I should only feel death for a moment. I'm sure I won't feel it afterwards. But what have we done to these Germans to deserve treatment like this?' hissed Fred.

He tried to control his fear as best he could. He knew it was unfair of him to crack up on Lily. Rarely was hard liquor kept in their house, even when they could afford it, but Fred remembered a few times when it had come in handy. Then he'd ingested every substance he could get his hands on.

'Thank Christ that one's missed us,' he would whimper as he took another swig, and Lily would try to reassure him with a cuddle. As they sat in the gloom and worried about their only son, flying ops over hostile Germany, they wondered if they would ever see him or another sunrise again.

Alternating scents of hope and despair permeated the air they breathed. No doubt a task worthy of St Jude, the patron saint of lost causes and hopeless cases.

He who foresees calamities, suffers them twice over...(Bishop Porteous)

CHAPTER 33

AWOL

Bill didn't like Germans but he didn't want to murder them. He didn't want to kill anyone for any reason, except perhaps some of the never-ending daisy chain of bureaucratic swine.

'All they're good for is to pass paperwork around. One giant merry-go-round of red tape and bullshit,' raved Bill. 'If they work slow, lie and say they can't cope because they're short-staffed, the pricks get more staff. If their department grows, they become more important and go to a higher pay level. Am I the only one who can see the hopelessness of all this?'

Sick of bombing Germany he wished innocent people could stop being maimed and killed in wartime. 'Major conflicts should be settled by placing the powers in one enormous arena to settle their disputes gladiator style. We could give them big sticks with gigantic nails protruding from one end,' laughed Bill.

'What would that achieve?' asked Alice.

'Nothing! But we could derive great pleasure watching those arse-holes inflict pain and suffering on each other, Alice.'

'Do you think it would ever come to that?'

'Probably not. There'd be some fast and fancy talking by scumbag politicians and suddenly no grounds for conflict whatsoever.'

'Certainly none of them would be around any big nails in the ends of heavy bits of wood for long,' laughed Alice.

Fred put down his newspaper. 'You're right though, Son. If ever a supreme power like a dictator, insists they settle disputes that way, there'd never be any controversy or conflict,' agreed Fred.

'To the survivor goes the spoils, including as a sign of some respect, everyone learns their language, salutes their flag and recites words of their anthem. What could be fairer than that?' asked Bill.

'The loser apologises because they lost, and cuts the best deal they can. Logical, Son. No more wars!'

'A tragedy of that magnitude has to be somebody's fault though.'

'Someone has to be seen to pay, yes. But not too much.'

'Verbal diarrhoea always flows in a downward direction, Son.'

'After those at the top have made it a dick measuring contest without equal, everyone becomes grudging but firm friends and trading partners again. Instead the conqueror spends horrendous sums of their own tax payers' money rebuilding the country they have destroyed and everyone is left to wonder, in a somewhat bewildered and bemused manner, remind me, was it us or the Germans who won that war?'

'Always remember, Kings may forgive each other, Son, but their people never can,' Fred said.

Bill disliked politicians with a passion. He labelled them as liars and thieves and accused them of groping in his trouser pockets for his small change. Such was his disgust for these dishonourable men he marvelled at their honourable prefix.

'I sense their fingers as they grope and dig deep around my genitalia,' said Bill.

Unhappy as he was with this Armageddon, Bill didn't return to Base by due time after his next leave. He and Alice had lost themselves for a few illicit days between a farmhouse room and 8 Barlby Road. Bill had learned to live from day to day. Every penny he had, they spent. Their only uncertainty was about tomorrow.

It wasn't long and Bill ran out of money, only then did he become concerned at the consequences of his actions. He decided to return to his Base at Syerston in Nottinghamshire. By now he was well and truly absent without leave.

At first no one gave him a second look. Streets were empty of Military Police until he strolled into the train station. Discovered with an out of date pass while he attempted to board a train from London he was chased down by MPs along a railway platform. Other service men deliberately obstructed them. His team players threw their kit bags in front of the enthusiastic pursuers as they gave chase. But Bill soon ran out of assistance and platform space. No more room to run and no other options meant he was apprehended, arrested and detained.

Detaining Bill meant the powers needed another Warrant Officer First Class or NCO Flight Sergeant to be with him on constant twenty-four hour detachment.

Bill couldn't be placed under close house arrest without another of equal rank with him. This would be for the duration or at least until his trial, court martial, or firing squad, unless His Majesty could think up an even worse outcome for him.

Bill wasn't an officer, but he was a pilot. The RAF was in a dilemma. Short of pilots, and with massive casualty rates, it was decided to drop charges against him and get Bill back into the air as quickly as possible.

For a moment Bill thought these double snouted bureaucrats had seen the light. The odds of his airborne survival were less than one-in-two. At least if he was killed, their problem would go away.

Alice was distraught. She preferred to think, rather than being a criminal, as a life- time sufferer of piles, her husband sitting on clusters of itchy and swollen varicose veins had caused his judgment to become impaired.

All was forgiven. The RAF, in their anxiety to get Bill and his crew back into the air with an irregular load of bombs, dropped all charges against him.

When Bill reported to Base he understood their urgency. Another major op was about to go down and he and his crew were listed. Snide comments between officers-in-charge were overheard. 'This should be a one-way ticket for a few. Let's hope a certain someone doesn't make it back!'

The op itself was the usual, pretty much what they'd come to expect. But the low level bombing height of 7,000 feet and grouping up at a pre-arranged assembly place was not. A three-quarter moon without clouds meant visibility was so good Bill could clearly see other Lancs near-by.

They were supposed to orbit a coloured marker.

Ben became tight lipped. 'Aye, the word is Skipper, it might be a pretty yellow.'

'Doesn't matter what colour it is,' groaned Bill. 'It could be sky-blue pink. Who cares? But it's another bloody beacon as to where we are.'

Joe had been pessimistic even before they took off. 'Their marker will attract enemy fighters from miles around. We'll be sitting ducks going around and around waiting for those clowns to call us in.'

Bill suspected Joe was spot on the money. He'd already decided to clear off about ten miles north, listen and wait until they were called in.

For a while they concentrated on their jobs; watched their airspeed, gauges, and the state of fuel. Bill saw in the distance aircraft going down. Some fell away after being hit without even a trail of smoke. Others exploded with a vivid flash. Bill looked at Joe. Neither of them felt at all happy with this waiting around for a Master Bomber to call the shots.

Joe clambered up into their astrodome. When he reported back his facial expression was pained. 'They're going down in all directions, Bill. I've counted ten so far.' His voice sounded strained with signs of panic.

Bill too became uneasy. Somewhere in the distance he saw a large puff of black smoke. Then another, each followed by an angry sheet of scarlet flame. He cleared off a further ten miles north, about twenty miles off target. It looked like being the usual RAF cock-up. At times like this he didn't care so much about the target, as not being one himself.

Another crew broke radio silence, 'Master Bomber, permission to come in and bomb.'

The reply, 'Denied! No, do not bomb. I repeat. Do not bomb! Continue to circle the marker.'

There was little chatter. Bill supposed like him, everyone was too busy trying to stay alive.

The same question not five minutes later. Same response.

Bill climbed to 12,000 feet and moved further away. To enter and orbit anywhere near that marker seemed like virtual suicide. Tracer was everywhere, most of it German.

Lancs already there at the designated height under light of the moon were going down at an alarming rate.

Joe counted another five, 'What's wrong with this clown, Master Bomber?' shouted Joe. 'Whose fuckin' side is he on?'

Another crewmember, a cultured English voice this time, spoke up clear enough to hear over the radio, 'Hello, Master Bomber. Pull your finger out of your arse. We're fucking dying out here.'

Then in a colonial accent, 'Fuck you, Master Bomber. We're coming in!'

Bill made his decision. He was no longer prepared to wait. He went in.

And so unfortunately did everyone else. Amid mass confusion, streams of flak coming up and tracer criss-crossing the sky, Lancs arrived from every direction in a hurry to drop their bombs and get away. Being further away Bill would be one of the last to arrive.

'Bill to Sandy.'

'Yes, Bill?'

'Sandy, you'll get one crack at this at about 12,000 feet. To hell with their 7,000 feet, I'm not going around again. You hear! I repeat I'm not going around again. Drop your fucking bombs as best you can.'

A great cheer went up throughout the aircraft.

'Well done, Arsey Bill.'

They continued on their bomb run and dropped their load. Below Bill the flak looked intense and accurate as more Lancs went down. He swung west and climbed.

The German defences had seemed well prepared. Bill gritted his teeth and flew on through turbulence. Flak exploded everywhere around him at all heights.

Bill felt under siege, and then a panic call from Frank.

'Hey, Bill, there's something coming up behind us. You see him, Bazza?'

'Bazza to Bill. Fighter to port upper. Prepare to corkscrew.'

Port upper. Oh, fuck! Did he mean to veer to starboard?

To his right appeared all clear. No other Lanc near him or below that he could see.

'Corkscrew to starboard, Bill. Go now!' yelled Frank.

Bill dropped his right wing, kicked the rudder and dropped his nose. As the great plane began to fall out of the sky he kicked it into a spin. This was what FOB had trained him to do.

When he straightened out there was no sign of the fighter so he must have gone off to chase easier prey.

'Bill, the wind is too strong from the west. Lay the wee plane off another ten degrees,' cut in Ben.

Then after about five minutes, 'Bill, lay the wee plane off another five degrees.'

Back down to maintaining a height of less than 5,000 feet Bill continued to fly over blacked out miles. Being caught from below by a German night fighter was not the way he wished to end up. Less likely the closer he was to the ground. When he could he preferred to fly at under 1,000 feet. He liked about 500 to 600 feet. Lower down it was more difficult to see him coming. Trouble was flying too low unsettled the crew, as there was insufficient height to bale out. It also gave 'friendly' guns less time to identify them so Bill gained height when approaching home ground.

Precise navigation by dead reckoning was not easy, nor finding a turning point in darkness or perhaps nearer home over water.

Suddenly, the blinding glare of a searchlight illuminated their cockpit. It hovered, then another, and another. They were coning him. Then came the flak. He felt a loud bang and an engine became irregular.

'Feather port outer, Joe!' yelled Bill.

The propeller wind milled for a while then prepared to stop as Joe feathered it.

They flew on with three engines. Ben constantly changed course. Bill hoped he might not see another enemy fighter or them him. He now limped, injured, but he should be near the coast. They swung west across the North Sea and Bill noticed one temperature gauge was into the red zone. Joe was ahead of him.

'It's beyond red, Bill.'

Bill considered his options. The temperature needle had vanished into an area past the threshold of where temperature needles were meant to go. That couldn't be good.

The rest of their journey home was uneventful although they were flying over water on their home run. With three engines and one of them running too hot, they were now committed to flying low across East Anglia.

'Bill, we're barely above the trees,' whined Sandy.

As they prepared to land, another problem emerged. Their starboard wheel lowered but didn't register locked. No warning light to confirm it. Joe tried to raise the wheels to attempt it again but the wheel refused to budge.

Coming in to land Bill thought, thank Christ we've dropped the bombs without a hang up.

It wasn't usually landing that bothered Bill. Taking-off was far more dangerous but this was different. It was all he and Joe could do to hold their plane. Elevator controls were not as responsive as normal but with careful use of throttles and flaps and handling the Lanc as gingerly as possible, he touched down.

Thoughts raced through Bill's head. He'd survived ops over Germany, been shot up more than once. Now he was skewing sideways crab-like down the runway.

It had been a normal touchdown if a bit fast before hell broke loose. His big fear that their undercarriage might cave-in had not eventuated. Braking hard didn't help, either. Runway lights in their path were smashed to pieces.

'Fuck, oh, fuck,' Bill muttered as the big plane ploughed off the runway across the grass. It was a critical moment. He had no idea how to fix it.

Redesigning the landscape was not intended. They never felt much of a bump when they passed through the first fence, although a few sheep fertilising the field might have looked surprised. They ploughed on through a few minor obstacles including the weather hut and came to rest in an adjoining field. There they stopped short of demolishing a house. Earth and dirt ploughed up from the bomb-aimers' compartment to meet them. Good job Sandy had vacated, thought Bill. 'Thank Christ we didn't end up in bloody church.'

Joe looked confused. 'The undercarriage is severely fucked up, Bill. But I'm not sure why.'

They had no idea what had caused the plane to misbehave so badly on landing. Then they looked closer and the reason was clear. Their Lanc had blown a tyre.

Landing a large aircraft like a Lancaster intact with a burst tyre was a virtual impossibility. They found all tyres showed slippage beyond allowable creep marks on rims.

'Another case of lack of servicing,' said Joe, 'I'm sorry, Bill. I never thought to check the bloody tyres.'

'It's no more your fault than mine. I missed it too.'

They prepared for de-briefing at some ungodly hour, a welcome home meal, drinks and bed.

Next day Bill complained to the powers about lack of servicing and maintenance.

'You're making a song and dance about nothing, Flight Sergeant. Watch yourself. It could be seen you're inventing problems to avoid flying.'

'But I did fly,' argued Bill. 'What I'm telling you are facts after the event, not before it.'

Bill was getting nowhere.

Joe and Bill had developed a unique relationship of mutual respect but they disagreed over issues such as this. Whereas Joe tended to blow smoke up the arse of anyone who outranked him, Bill saw it as 'crawling,' something he had difficulty coming to terms with.

Alice thought it understandable and fair as his life was on the line but the same applied to others, as with Joe. His life was on the line too. Joe might see the 'game' being played as less draconian but to Bill's mind there was no room for error. If they were ever going to survive this war, changes needed to be made. Shrapnel might have contributed to their blown tyre. Or perhaps lack of maintenance. All that weight careering out of control, brakes non-existent or pulling the wrong way, one wing dipped, engines out of balance, it was hard to tell, which was the greater culprit.

More fuel for Bill's recurring nightmares to come, would now be a corpse-strewn runway.

Overnight the plane was left in the paddock.

'Pity the wee plane canna stay in the paddock permanently,' said Ben.

Fear, guilt, panic and despair meant the rest of their day was nothing special to remember.

If you don't like something change it. If you can't change it, change the way you think about it...(Mary Engelbreit)

CHAPTER 34

LMF

Weather conditions had been harsh with crews stood down. It was a welcome break between bombing Berlin. The big news was bandleader Glenn Miller had disappeared on a flight from England to Paris.

At Base some said his pilot took him through a no fly area, across sea designated for ordinance disposal. Bill was unconcerned. For him it was just another day at the office.

'What the hell! After a forty-eight hour leave, we've another mission. For the lucky ones there's always another op.'

In their Sergeant's Mess Bill excused himself from his crew to go have a drink with Olly, who was now an officer.

'You slumming it, Olly? You shouldn't be in here with us NCO's. Why aren't you in your officers' bar?'

'Aye, I wanted to have one with you, Bill. I'm as dry as an Arab's jockstrap. It was summat to invite you in there, or me come in here, before our next op. Seems a long time ago, doesn't it?'

'What?'

'Canada of course, Bill. Canada.'

'It certainly does, Olly. But as you know, I'm a married man now. I've got Alice. There are aspects of Canada for me that are best forgotten.'

Olly called their steward to buy some Woodbines.

'Sorry, Sir, we're out of stock.'

'Aye, all right I'll have some Players No. 3 then.'

Their steward looked embarrassed and coloured when he said, 'I'm sorry, Sir, but we're out of stock of them as well. A consignment should arrive tomorrow or day after. Would you like me to put some aside for you, Sir?'

'Aye, put me down for some,' answered Olly.

Bill felt a wave of disbelief. He would never have tempted fate to assume he might be back to collect an order.

Bill shook his head. 'Very brave of you.'

'Aye, she'll be right. We're both alive but I understand where you're coming from. Naught such concerns for me superstitious wise or woman wise. As you know I've never married but your Alice, she's a fine young woman. Aye, you're a lucky man, Bill.'

'Yes, I've wondered about that, Olly. You not being married, there was a time there, when I thought it might get serious between you and Anne, but then it never happened and she was killed. I suppose we don't ever get over something as tragic as that.'

Olly coloured. 'Aye, thousands of innocent people are dying on any given day, you mean? And now, you want to get sentimental or summat over a nurse who was naught but a whore!' Olly had raised his voice in a maniacal way. A few drinkers at other tables glanced their way anticipating a quarrel.

'You feel strongly, Olly. Perhaps I shouldn't have mentioned Anne. I'm sorry.'

Olly calmed down. 'Aye, and by gum, Bill, I'm sorry too. I've been carrying naught but summat of a burden about it, that's all. Richard seemed able to put our church hall antics behind him, but to me, it was so real. Aye, it was like living a dream with Anne. She already had Richard as you know but she was my girlfriend for three whole days prior to the wheat field incident with Jill. She was a bloody tiger, by gum. She nearly took me back.'

Olly's voice trailed.

'You liked Anne, then.' Bill felt embarrassed to mention her name in case Olly sounded off again.

'Aye, I naught but liked Anne. I loved her. I loved that woman summat bad, so bad it hurt. I could never get her out of my mind. Then again, I liked Richard, loved him too in a different sort of way. But Anne didn't feel about me the same way I felt about her. She was promiscuous and a bit carried away with sex with other blokes. She bedded Richard, then me. And who knows how many others? How can a man marry a woman like that?'

'It would make for an exciting life, Olly but to be honest, a bit too adventurous for me. What do you do when you come home every night? Get down and sniff her pussy to see if another man's been there while you've been at work?'

'Aye, maybe Bill, but there's summat you don't know. Within seconds Anne could change into a harridan and belittle. It's time I got things off my chest. You're the only person I feel right telling, by gum.'

'You can tell me anything, Olly. You should know that. You and I have been in this madness together since day one. There shouldn't be anything we can't confide.'

'Aye, tomorrow then, Bill. After we return from our op. I'll spill my guts, summat in your ear about Anne. Will that be okay?'

'Yes, of course it will. But you're making me feel uncomfortable, and you know how I feel about pledges before an op. Neither of us might make it back.'

Olly rose to leave. 'You'll be right, Bill. Aye, I guarantee it. And to prove it I'll watch your tail.' He laughed then drained his glass and left.

Olly turned at the door and gave Bill a wave of his hand. Then he was gone.

Bill felt ill. It was as if Olly's wave had been Richard's. Déjà vu.

It was the usual briefing. It rambled on, all the repetitive crap. Bill was aware he should pay more attention. When he did listen, most went in one ear and out the other. Their op was later that evening; a Saturday in October 1944. It was to be an all out assault on Germany's large synthetic oil and benzol supplies in the Ruhr Valley. Their target was well to the east and required long-range fuel tanks to be used. More weight! Don't you fuckers ever learn? All up with every squadron involved it was tipped to be a biggy of more than 1,000 Lancs.

After their propaganda meal Bill and Joe prepared to carry out their pre-flight checks but they were unhappy with two engines. Their sound was different.

Synchronised engines had a note about them. They almost sang. But on closer inspection two of their four Merlin engines appeared to be well below par.

'Look, Joe, when we rev them up, two oil lights remain on.'

Joe was concerned. 'Doesn't sound at all right, Bill.'

They discussed and agreed.

'Port inner and outer sound rough.'

'Oh, fuck! Bill. That's two engines on the same wing.'

'Port outer has a more distinct knocking sound and look, there's glycol streaming out of it.'

The dismal overcast day matched their mood. Frank stuck his head around into the cockpit. He made fun of pushing aside myriads of oxygen hoses and intercom leads that became easily tangled then made an announcement.

'My mid-upper gun turret is rotating slower than usual, Bill.'

'What the fuck do you mean?' asked Joe. He and Bill looked at each other as if trying to decipher a puzzle.

'Well, it's slow. And its groaning noise sounds like someone being hit with a bloody big stick,' grinned Frank.

Bill called for a mechanic.

On review, he scratched his head. 'I don't know much about these Mark One engines. I was a waiter before I joined up, Sir.'

'Bloody encouraging, mate,' said Bazza. His tail turret worked perfectly but he was pleased to be included in any shit-stir.

'Maybe the engines need a major service,' suggested the mechanic.

'Maybe won't cut it, Corp. No good asking us, we're asking you,' said Bill.

'They changeover Rolls Royce engines on Spitfires after every eighty hours of flying,' said the mechanic. 'But not these.'

Bill let out a low whistle.

'Then these are well overdue,' said Joe.

Had their adventurous landing buggered up the warning lights? Bill was worried enough to request further mechanical attention. 'Get someone to look into this, Corp.'

'Preferably someone who knows about Merlin engines, not someone from the local pub, practised at pouring a pint,' reinforced Joe. 'What we don't want is to blow two sets of bearings over Germany.'

As other Lancs prepared for their take-off instructions, Bill stood his crew down.

Within twenty minutes he was summoned to the CO. He sat in comfort behind a large desk while sipping a cup of tea.

'Managed to get one not laced with bromide did you—Sir?' asked Bill sarcastically.

Bill's frivolous question was ignored. 'Flight, the plane is perfectly all right to be flown.'

'Good—Sir. You fly it then.'

One thing led to another. Bill argued.

'The plane is faulty—Sir. That aircraft requires a refit of two engines before it drops bombs on Germany again.'

Final outcome, Bill refused to take the plane up. 'It's one thing for a Lanc to limp home on less than a full complement of engines without its bombs. But it's totally different if we have a malfunction on the way out with a full payload and only two engines. How will I gain height?'

'If you know what's good for you, Flight, you'll take her up without any further delay.'

'Added to which mid upper turret won't budge,' argued Bill.

'If you don't get a move on you'll be too late to join the main wave!'

Bill came under considerable pressure to fly his aircraft on that op. But in his mind he'd reached a crossroad.

Was it better to be a live rebel than a dead 'yes' man?

Suave yes men seem popular with the RAF, thought Bill. But surely not for long? They're unprincipled and their actions predictable.

This posed another interesting conundrum for the RAF. Perhaps they were still smarting from letting Bill escape from the AWOL debacle? Planes and pilots were in short supply. Bill knew this and by standing himself and his crew down, they would miss this op.

He understood it could be seen he was making hollow excuses, unwilling to obey a lawful command to fly that big bird. It wasn't difficult for the powers to deduce Bill would rather be in Alice's welcoming arms than dropping a load for King and Country over Germany.

'Does your wife still live down the road?' asked the CO. He looked about ready to carry Bill over the abyss.

'What's that got to do with it?' argued Bill. On this occasion, if proven, he would be guilty of a lot more than being AWOL.

LMF was unlikely to be overlooked. He was literally one step away from being frog marched away to the guardhouse by two military policemen and his locker cleaned out within the hour. If this went badly for Bill he would be dishonourably discharged. He could finish this war not as a pilot, possibly not even peeling potatoes in the navy, but most likely in jail. These were the most serious charges he could have dreamed up for himself and carried a high price if those engines were proven to be in good working order.

Even if he somehow avoided jail, dishonour and shame would haunt him for the rest of his days. He was justifiably concerned.

'I'll give you one more chance and only one,' said the CO, neither friendly nor hostile. He rose from his seat and glared at Bill. 'Take that Lanc up now. Do your bloody duty, coward—or face the consequences of your actions. Is that understood?'

Part of Bill wanted to toady and say, 'If you knit me some wings, I'll fly the fucker in any colour.' But he remained adamant despite his lack of mechanical expertise.

'Why have oil warning lights if they can be so quickly disregarded?' whined Bill. 'And what about all that coolant pissing out?'

'You have my instructions, Flight.'

For a moment Bill wavered. It would be so much easier to follow the order but something bubbled up inside him. There were no extenuating circumstances or shades of grey. Olly came to mind. He'd told him if push came to shove, he was the pilot. It was his decision. Otherwise to Bill, that made him a hypocrite. *Shit!* Last time he felt like this was when challenged by Alf and his mates. This was different, part monster, part magic but still bullying. He could also lose more than his teeth.

Suddenly this negotiation had assumed the importance of the Treaty of Versailles. Tears of white-hot rage welled up inside him as again his sense of right from wrong never wavered.

Fuck them and their plane!

Bill straightened almost to attention.

'My decision, is to stand my crew down—Sir.'

'Very well. It's a great pity you have chosen to take this path and not let the matter drop. You were a promising pilot. Senior mechanics will examine those engines, throughout the night if necessary. And, when they're found to be all right, you'll be charged with Lack of Moral Fibre.'

Bill was defiant.

'Go to hell—Sir.'

'Dismissed!'

Bill stuck with his decision not to take that thirty-ton birdie up. His crew were delighted. It wasn't their call but his. Then Bill started to worry. Now a large noose dangled as he moved effortlessly towards it.

The other Lancasters had already taken off, gone on their op when Bill retired to the otherwise empty Mess with his crew.

An officer found him there and he looked angry.

'You haven't heard the last of this, Flight. You'd better watch out. You're more than a little bit fucked!'

Joe let out a low whistle. 'They really are pissed off, Bill.'

The rest of the crew kept their heads down and kept drinking but Bill was not in a thirsty mood. He retired early and stayed at Base. Bed was where he tried to find sanctuary but throughout the night he tossed and turned. His bedsprings growled and moaned as though in mortal anguish. They'd sung their discordant tune before but never like this night.

The constant reminder of his piles didn't help. He couldn't sleep with the enormity of it all, then when he dropped off, nightmare followed nightmare. Utter hopelessness intertwined with despair. He couldn't escape hideous visions of himself within and without in three dimensions. In his dreams Alice was drenched in tears. He'd never seen her cry like that.

He went from being incarcerated in chains in various dingy water filled dungeons to execution. Not once but numerous times. First he faced a firing squad, then decapitation with a blunt axe. There was a forest of nooses. Then he was hanged on each. Past the point of exhaustion by morning far from feeling refreshed, he felt drained. And he didn't feel nearly as firm in his conviction as he first thought.

Up early, he was shaved, dressed, piles taken care of, and with the definite 'hang dog' look of a convicted man, trudged over to the hangar. There to see firsthand the consequences of his actions. And perhaps get the heads-up, as to which dungeon might have one small barred window.

Fuck! The hangar was completely empty.

No mechanics and no Lanc. His Lancaster bomber had gone!

Bewildered, Bill wrestled with his worries alone. Then with a heavy heart he slunk like a lamb to the slaughter to his CO's Office to hear the verdict of what was to become of him.

'Reporting for duty as instructed, Sir,' said Bill, with a half-hearted sloppy salute.

The CO looked first at his own shoes, then at Bill's. Slowly his eyes travelled up Bill's body but stopped at his neck. Their eyes never met.

Bill blinked and when he opened his eyes the man opposite swam into view.

He's a slippery sod at best of times, thought Bill. But he lacked his air of cocky supremacy this morning. His confidence had changed, replaced by awkwardness as he stared down at his papers. Bill noticed his lips trembled and his face was pale. He remained almost motionless, except for constant clasping and unclasping of his sweaty hands.

'Oh, yes, that's all right, Flight—at ease! You're dismissed.'

For a moment Bill stood and blinked.

'Dismissed, Sir? What about the Lanc? What's happened?'

'Oh, that, you're not to concern yourself. Everything's in order your end. You can stand down.'

It was the final straw. When would this demoralisation be complete? Would lies and corruption never stop? There had been prevarication, politics, incompetence, deceptions and now to top it off, his loss of sleep. Unnecessary at the hands of these total arse-holes. Bill's anger and frustration yawned big and hungry in his chest. He wanted to strike out, but knew that might take his protest of hurt indignation too far. Instead with a bellow of rage that would have rattled the CO's curtains had he had any and been heard in Calais, Bill exploded.

'You're lower than low. You're a disgrace to the men who serve.'

'Watch your mouth, Flight. Anyway, low is a relative term.' Only point of reference was his darting eyes.

'If you want to help the Germans you're going the right way about it!' shouted Bill.

'No comment.'

'What! No comment? No fucking comment? What's that if it isn't the last refuge of the truly screwed!'

He was angry, wanted to be as ruthless as Lily when she snapped pigeon's necks for their table. Instead he turned and stormed out. They could add insubordination to his list of miscreants but he didn't care. He'd had a bad week. He went looking for Olly.

At the de-briefing he'd missed there had been almost a full complement from their squadron, but one crew in particular he was told, had never made it back.

Bill was devastated to find Olly's crew had not returned.

His friend gone, Bill had that awful bitter taste in his mouth again.

'Might he have baled out? Joe asked.

'Aye, there's always a chance. Perhaps they've survived,' said Ben.

Thoughts of Olly about to be interned as a POW for the remainder of the war buoyed Bill's spirits. But others had seen a Lanc hit.

'It burned up and went down without any sign of survivors,' they said.

Olly was listed KIA—Killed In Action.

Bill wanted to take up the cudgel for his friend. To shout, 'you've killed another decent man and his crew. You uncaring sods!' But he didn't. He knew even if he had, he could never substantiate his claim. He felt a great sadness and a welling anger as he fought hard to control his emotions. It reminded him of when he'd been told Richard's plane was missing.

This was different. If Richard had perished it was by his own hand, pilot error. Perhaps he ran out of fuel, got lost and crashed. This was far worse. This was Olly. He had known Olly for longer than Richard. They'd trained together, been assigned together, and been combat pilots together. Saddened, Bill was distraught at the loss.

Thoughts of never seeing his friend again nor watch him laugh and say, 'Aye, Bill,' and 'by gum, Bill,' un-nerved Bill. It also occurred to him, whatever his friend Olly was intending to tell him about Anne would never be told.

The carton of Players No. 3 might remain in the bar until someone else wanted them. Meanwhile another crew of young men had been hurled into the cauldron of loss and despair. Irrespective the RAF always insisted Lancasters were safe to fly. It was a case of, 'certainly we expect injuries and death, but nothing too serious.'

Bill discovered through grapevine gossip, a Lancaster bomber had indeed mysteriously disappeared off Base in the middle of the night. No registration, no numbers, nothing was filed on any paperwork. Its destination was supposed to be a secret. Apparently it had been the only movement out in twenty-four hours besides those that went on the op.

Not unusual to fly an op but an op, thought Bill, but for only one Lanc? That was rare all right, so rare as to be unlikely. Then word was it had been transferred to another Squadron. RAF Gransden Lodge was muted. But to move it under cover of night and in total secrecy, why? Bill couldn't believe it.

'Somewhere with better facilities for re-building engines,' was touted.

'But Gransden Lodge has no better facilities than here,' ranted Bill. 'And if the Lanc was considered safe to be flown, then why would any engines need to be re-built?'

Bill asked a lot of questions. He demanded answers and wanted to know where the bloody hell was his thirty-ton birdie?

The official line remained, 'Look, Flight, we're too busy with the war. Sorry, end of story.'

Bill's crew agreed. They were of the opinion two of those four engines were defective. Such an extensive rebuild was unlikely at that juncture of the war. Planes and pilots were undeniably in short supply.

That the war effort would have been better served had proper maintenance been carried out in the first place, and then everyone could serve the same cause was a given.

'I'm convinced, Alice, that Lanc was reassigned to another active Squadron—like Gransden Lodge. Somewhere perhaps where another young pilot might be less discerning, less outspoken. More committed to toe the line and accept pressures applied to fly that bloody plane.'

'Oh, Bill, could they be that cruel?'

Bill thought for a moment. 'Yes, I believe they are.' In that moment, Bill knew he had his answer. 'With the RAF there is no right or wrong. Only their way.'

The obvious outcome in Bill's mind was more than catastrophic for the poor unsuspecting crew assigned to fly that limping dodo. Bill searched his soul for answers. He knew 'fairness' and 'fair play' had never been part of the RAF's over-riding considerations. But this! This was tantamount to murder. 'I didn't handle it at all well, Alice,' said Bill.

Alice was quiet and then she said. 'But you did handle it, Bill.'

'True, Alice.'

Bill faced another example of his reaffirmed disrespect for the hierarchy of the Royal Air Force. A further annoyance festered in his brain.

'They might be good at ceremonial occasions, and they might be quick to demonstrate publicly their humanitarian side. But when it comes down to the basics, like an inquiry of their own military disasters, its record is consistent all right, as in not looking too hard for fear of disclosing its own incompetence. Their whitewash, their lies, gross corruption, and general lack of ability in higher ranking officers, when kids' lives are being lost and not their own, sickens me to my stomach,' said Bill.

'The usual bugger's muddle, Son,' said Fred. 'Nothing's changed. It was the same in the last war, my war.'

'Have a nice cuppa tea, both of you,' Lily suggested. 'It's never quite so bad after a hot cuppa.'

Outside, fat raindrops dappled pale new leaves.

Nothing improves the memory more than trying to forget...(Anon.)

Chapter 35

SHAKING OFF NOTTING HILL

Alice's family seemed content to remain in Notting Hill. Unlike Alice and Bill they showed no misgivings about their poor surroundings. Nor did they appear to want anything better. This created rifts between Bill and them, but as Alice was committed now to Bill it didn't matter much to her.

'I don't think they'd move if touched by a red-hot poker,' quipped Alice.

'Never mind them. They're like prisoners. Content to hug their chains.'

Alice was embarrassed when Bill visited what had been her Notting Hill lodgings. To save money, she'd had a candle stub beside her bed. Bill stared around at the pockmarked walls and gas lamps. Christ, they haven't even electricity, he thought. Her brothers shaved in the reflection of a piece of broken mirror propped up over a bowl. Their sparse and battered furniture, prison-cell type beds and faded wallpapers etched into damp walls. The small ill lit rooms smelled like a homeless wet dog.

Bill's visits did nothing to heighten his opinion of Alice's family.

'At least Lily and Fred agreed to move out of this shit-hole, Alice.'

Another thing; Bill smoked a pipe but heavy cigarette smokers had carrot-coloured fingertips and ingrained yellow teeth. That is if they had any teeth at all. Dentures were the norm in Notting Hill.

Bill had grown to dislike gruff working class manners. He swore like a farm hand but disliked their working class, unwashed look. Men wore wide leather belts and braces over their shirts without collars, and top buttons of their vests always showed above their shirt at the front. Women seemed to spend their entire days in curlers or with cheap scarves over their heads drawn up at the sides. They walked the streets in battered carpet slippers.

When Bill accompanied Alice on visits, he sat most of the time not wanting to be there. He tried not to sit in anything 'yucky' and watched the trains from a rear window. Waiting for a steam train to pass became the highlight of his day.

'Even their tea, their grouts, and their tinned milk served in cracked cups is distasteful.'

Alice agreed.

Lily and Fred, Bill accepted, were a cut above. He'd never noticed before but he could see it now. Had he become a snob?

Alice saw less and less of her mum, Fanny. It upset her because she had a high regard for what Fanny had achieved with bringing up their family without a man about the house.

Fanny was a remarkable down to earth woman who displayed an unusually high standard of moral code, which she passed on to, perhaps not all of her daughters, but certainly to Alice.

One of Alice's sisters was Marge, married to a drayman called Jack, who worked at the local brewery. They lived in a council tenement near Hammersmith but Marge, while married to Jack, had played up. She had sex with the coalman on her own front door step.

'Goes to prove,' said Bill, when Alice told him. 'The knee trembler has been around for a while.'

Bill hoped for Marge's sake the coalman fared better than he had with Sophie.

As the result of their clandestine meeting was illegitimate child, Harry. So, obviously he had.

Jack was a giant of a man. Bill watched spellbound during a sisterly visit with Alice, as this human bear dunked an entire digestive biscuit into his tea and then into his mouth in one piece. Jack was what Lily and Fred called a straight man, and he offered to take baby Harry in.

'I'll bring him up as one of our own.'

Not an uncharitable gesture and according to family gossip, an excellent upbringing for young Harry. But Marge wanted to get shot of the baby, to get him out of her way.

Before modern adoption societies with regulations, rules and licenses, which would arrive in due course, Fanny stepped forward and took the boy in as her child.

Her attitude was simple. 'Charity begins at home, Marge. If you won't look after your own, then no grandchild of mine is going anywhere, but staying home where he belongs,' said Fanny. Here was a woman, with no husband, who had raised nine children of her own, albeit now grown up, who worked charring as she'd no other means of support. Yet she took in little Harry because she considered that was the right thing to do.

Alice had never known her dad. She knew she had one but didn't even know his name. Fanny told her, 'When I was pregnant with you, Alice, your dad said, 'enough is enough, Fanny. I'm off.'

Alice was upset about that while growing up. She believed it was her fault her dad had left her mum. 'Not at all, Alice,' said Fanny. 'The relationship staggered on. It got bleaker by the day. He was no good, your dad. We were better off without the likes of him.'

Fanny had a brother, Ned, who lived next door at number 88. Ned had fought in the First World War and been injured. Shrapnel in his guts had upset his bowel arrangements. In hospital his intestinal tract looked like a scale model of a Hornby train set with bits of glass tubes running everywhere. He spent most of his days lying down or sitting, then needed to empty and clean out his tubes daily. Ned was in a bad way. When he could get hold of any cheap gin he drank it, as much as he could manage. But no one criticised him. They agreed he'd done it tough. If he could enjoy a little bit of pleasure if only for a short time, why not?

Shortly after the Germans started the Blitz, Ned had become quieter than usual. With his disabilities it was a long slow process getting dressed to try and get to an air-raid shelter. Most times when warnings sounded Ned stayed put. He sat in the dark alone and pondered his fate month by month as bombs dropped close by. His nerves stretched to breaking point. Then after an all clear one day he called Fanny.

'I can't take this anymore, Fanny. I'm sorry.'

Next day he forced himself to run the stairs, which shattered glass tubes in his guts. Ned died in absolute agony stretched out on a landing floor.

'Fuck me, Alice. Couldn't the poor bastard think of a better way to end it?'

Alice was upset and frightened. Every time she visited she tried to avoid looking at where Ned had died.

At Barlby Road on their mantelpiece was a framed photograph of Alice and Bill outside their Registry Office wedding. Bill looked smart in his RAF uniform, as did Alice in her unusual hat with a coat and dress to below her knee.

One day a pious visitor asked. 'Why did you marry in a Registry Office, rather than a church, Bill?'

'I drop bombs on masses of poor souls. Then I realised most likely they're praying to the same Deity, but harder as there are more of them.'

'Who is this Bomber Harris?' asked Alice.

'Air Chief Marshal Sir Arthur Harris is known as fiery Bomber Harris on account of his red hair, Alice. To me and my crew he's Butcher Harris. Not only is he determined to outdo Field Marshall Hermann Goering by dropping more bombs on Germany than thrown at England, but he's inconsiderate of RAF crews' lives,' answered Bill.

'Is he qualified?'

'He's not an academic. Comes from a family steeped in the tradition of serving the British Empire in India. Word is rather than attend Woolwich or Sandhurst for officer training, he elected to go to the British colony of Rhodesia in Southern Africa and become a farm manager. Years later he returned to London and through his family contacts pestered the War Office for a service commission with the Royal Flying Corps.'

'Then what?'

'He became noticed and eventually had an ally in Churchill who saw reason to agree with his ideas. In October 1944 Harris, who in my opinion had found the ideal way of going to war in a safe and sitting posture from Whitehall, instructed his CO's to bomb Duisburg, an industrial port on the Ruhr River. It was significant because the RAF sent more planes than ever before over a period of twenty-four hours.'

'Why is he Butcher Harris?'

'We nick-named him Butcher Harris because he claimed the raid was so successful similar ops should follow. Figures were mooted of sending 1,000 planes with losses of between ten and twenty percent deemed as acceptable. He and Churchill make a good fucking pair!'

If a man hasn't discovered something that he will die for, he isn't fit to live...(Martin Luther King Jnr)

'Doesn't mean he's entitled to send others to die for him, though,' said Bill.

Chapter 36

INSUBORDINATION

With many hours of combatant flying to his credit Bill had only a few ops left to complete his tour. He'd started to become hopeful he might survive.

That's a dangerous new strategy, he thought.

Being part of such large raids had the dubious advantage of being lost somewhere in the middle, as long as he didn't crash into another Lanc, or someone else blunder into him. Unfortunately that happened all too often.

Butcher Harris had problems convincing his bosses how his reckless strategies remained best policy and that his style of intensive night bombing, if continued, would win the war. There was doubt arising as statistics released showed more RAF aircrew had been lost over Germany than enemy civilians killed on the ground. This could no longer be sifted and regurgitated by Air Ministry to mean something else. Those who wanted to keep their jobs after the war began to panic.

Christmas approached without peace on earth.

'Goodwill towards men has become a bitter dream, Son,' announced Fred.

'It's on again in Belgium, Dad. German tanks and infantry have resumed their offensive.'

But that was about to change, as was Bill's strategy to stay alive. It had altered and continued to evolve.

'If nothing else, our Bill's fluid in his thinking, Girl.'

As long as he wasn't sitting directly underneath another Lanc he could see the advantage of being one of a crowd. It could mean a degree of safety in numbers. Might be something in that poster after all; don't be proud, stay with the crowd.

The key to Bill's new strategy lay in the awful plans of Butcher Harris. In a normal op being one of twenty or thirty Lancs was one thing. To be one of 1,000 planes, another.

After an op later that month the powers had another idea of self proclaimed brilliance. It was decided to check out current thinking of their more senior pilots in a sort of remedial test. Each pilot was instructed to report in full flying kit, presumably to remain in character, and then take turns at sitting in a hangar answering a myriad of questions.

The test was intended to determine how well they knew their piloting jobs.

'Great idea,' said Bill. 'Some high brass must have been bored shuffling papers, to think up this crazy crap.'

Their test comprised a string of bureaucratic questions in order to assess if pilots had the ability to succeed at what they were doing.

'Perhaps they should survey those who fail to return,' grumbled Bill.

The panel comprised three high-ranking officers. Bill was unimpressed.

'Big rings on their sleeves and covered in mountains of metallic egg yolk,' was how Bill described later the layer upon layer of their golden insignia. 'By the pound these boys were top notch. But they could also afford to be ruthless and aggressive bullies.'

There would be plenty of opportunities for any pilot interviewed to keep jumping up to attention, salute to perfection or even remain frozen in a permanent state of salute in the presence of such dignitaries.

Bill didn't like the idea of being questioned by these officers who were experienced at flying desks or at least a desk they'd been given. In addition he was supposed to be on a treasured day off with Alice, not playing silly buggers in full flying kit to a panel of three fuck-wits, trying to justify their existence. That none of them had ever flown a plane in combat further ratified Bill's unsavoury opinion of them.

'Jokers at the top of the totem pole, Alice,' reinforced Bill. In his opinion they lacked any 'real' flying experience. He was not content they enjoyed this safe little game at his expense and outranked him on a higher pay level while doing it. Bill made a conscious decision to be more difficult than usual. He didn't have the wit to stand up when they came in, to salute them, and other than being there, showed no respect whatsoever. He set out to rattle their cage but in hindsight, which is always 20:20 vision as he told Alice afterwards, 'my not being there might have better suited any promotional prospects.'

Questions were straight out of a basic flying manual and slanted to a foolish degree, hardly appropriate for grown men flying combat operations at that stage of the war.

They appeared preoccupied with knowledge about safety procedures, which incensed Bill as he considered the recent debacle over his two engines.

'Perhaps they were attempting to cover their arses by showing an otherwise invisible caring side,' sneered Bill. 'When all this is over, difficult questions might be asked.'

Quoting recommended safe pay loads were further examples of their stupidity and went a long way to reaffirm his original assessment of their poor social standing;

'What difference does that make, Sir?' Bill asked in response to a panel of blank faces. 'When I take off with bomb doors locked open I know the plane was never intended to transport the bomb loads we carry—Sir! We are often loaded to the point of the undercarriage buckling—Sir.'

The panel conferred. Not as if for pleasure in each other's company, more as if they wanted to be surrounded by their own courts, by men subservient to them. They chose to overlook his comment and wanted to know his optimum flying height.

'If I ever had a correct barometric pressure, or an accurate altimeter, I might know my correct height and I could tell you—Sir! But as the instrumentation is unreliable even when it is functioning—I can't—Sir!'

The panel pressed. They wanted some indication.

'Low—Sir—as fucking low as possible—Sir! So low Schrage Musik cannons your lot insist don't exist, can't get me—Sir!'

The panel frowned at Bill's potty mouth and said they would prefer he confine his answers to their questions.

He thought he had.

'Why do you use expletives?' said the man in the middle.

'Fucked if I know—Sir!'

They wanted to know if he drank alcohol in the mornings.

'Yes, Sir—every chance I get. Presuming I know it is morning. Pity there's none here now—Sir!'

They wanted to know if his drinking was a problem?

'Better since I gave up hope—Sir.'

They wanted to know what part of the flying experience he found most fulfilling.

'Coming home, as in landing—Sir!'

Yes, the panel understood he liked to get back safely. But they wanted more to do with his flying experience.

'Nineteen shillings and sixpence a day, it's a good incentive—Sir!'

Then Bill thought for a moment. 'On the rare occasions I fly in daylight, I particularly like climbing above the clouds where the sun is always warm and inviting, and the sky is a clear blue. But I don't stay there long—Sir!'

What about taking off, the panel wanted to know. 'How do you feel a Lancaster responds to your controls, Flight Sergeant?'

'Poorly, Sir! But if we ever carried bomb loads not well in excess of safety recommendations, perhaps I might know and could answer you more fully—Sir!'

They continued to push.

'The plane responds like a lazy cow. I often fly to Wales to get enough height. In my opinion, with an overfull payload the Lanc would be better suited with three Merlin engines strapped to each wing instead of two—Sir!'

They looked at him as though he was something they'd trod in before one final question. It was about a lever, which if activated, would jettison the bomb camera.

Bill was asked if he knew where the lever was situated inside the plane.

'I have no idea—Sir. But now I know I have one, I might give it a go next time I'm up—Sir!'

A ruddy-faced examiner was almost the colour of beetroot. It matched his purple nose and cantankerous demeanour, thought Bill. Laden with egg yolk his shoulders sagged under the weight of crapola. He wanted to know why Bill had not made it his business to find this lever, if he didn't know where it was?

'I don't give a toss where it is, as I have no intention of ever using it—Sir.'

All three became quite excited at this and wanted to probe their subject with more detailed questions. If the plane were in flames and going down, why wouldn't he want to jettison the valuable camera and its irreplaceable content?

'Absolute tripe—Sir! If the plane's on fire and going down—Sir, I've more to worry about than that camera—Sir. And if the plane's going to crash and we're going to die—Sir,I can't see a single logical reason to drop the camera in enemy territory or out into open sea—Sir. It can burn with me—I mean us—Sir!'

The panel continued to be unhappy with Bill's attitude and responses.

They whispered among themselves and made copious notes. The officer in the middle ventured to ask if Bill might like to examine the fuselage of a Lancaster more closely, to positively identify more accurately location of the lever in question.

'No, I wouldn't—Sir.' Bill was more than a little bit irate. 'And I strongly suggest you three should have something more constructive to do with your time too—Sirs. If you want to investigate anything at all, try investigating why a pilot is instructed to fly a Lancaster over laden with over-weight bombs, with two obviously faulty engines, indicated by oil lights remaining on, coolant streaking everywhere and no-one but me is even slightly interested in the welfare of my flight crew—Sir!'

Bill was on a roll.

'You're talking bollocks—Sirs! Your tiresome and unhealthy interest in this camera lever is tedious—Sir. If you have nothing more important to add, like why there was no enquiry into a pilot being coerced to fly an aircraft reported as faulty I disrespectfully request permission to get out of this uncomfortable flying kit and go see my wife before I fly my next op—Sir.'

'Flight Sergeant, you are dismissed.'

At home Bill told his story and realised too late he'd overplayed his hand.

'There are certain lows, Dad, I might never forget. Everything I told them was as true as if it was the devil's spit.'

'Things to know but better not to tell them, Son.'

Bill had failed to play them at their own game. His report on their downfalls had missed the verbal connection he required. His honesty harmed himself.

'Why did you bother, Son?' asked Lily.

'I'm not sure, Mum. Perhaps it was to make myself feel better. Maybe, I felt I owed it to Olly.'

'You can take the boy out of Notting Hill but can't take Notting Hill out of the man,' said Fred, as he shook his head.

In retrospect, it was easy to see where Bill's decline with the RAF began, but at the time it was more difficult. Even when he thought back, he would wonder at his own faulty memory.

'You'll never acquire the accent and manners of officer classes, Son,' said Fred.

*Silence is a friend who will never betray...(*Confucius)

CHAPTER 37

NCO

'You've a way with words, Bill,' said Joe.

'Aye, Bill,' agreed Ben, 'Naught doubt about it. You tell it the way it is, that's for sure.'

'Trouble is, way I see it you're never going to get promoted, Bill,' said Sandy.

'But I'll bet you never forget the sound of a knocking engine.'

Bill scratched his head and pondered. His star was not on high with the powers but despite his unpopularity with them, his crew had a high regard for him. Should he be prepared to sacrifice becoming an officer, provided he was well respected by his peers? Comments bandied about showed there was never a shortage of men prepared to put their lives on the line to fly with him, if need be, which tended to unsettle the powers. The record stated differently, and proved evidence of Skinny.

In addition to having a reputation for being a no-nonsense straight speaking man, Bill had the dubious distinction of being one of, if not, the more experienced pilots. But, while plain talking and straight dealing can be commendable traits, battering his superiors with the untempered truth had a vulgar bluntness about it.

Undaunted Bill continued to batter.

Partly due to his longer training in Canada, which had been discontinued due to pressures of supply, demand and time restraints, he was a proven survivor. Excluding his intoxicated plane-turning debacle, as an older hand new pilots often asked Bill's advice. He'd never given it a great deal of thought until now. Perhaps there was something in what was being said. He was a survivor, when other good men, better than him or not, had not returned.

As His Majesty George VI became more and more desperate for pilots, sixteen week training plans plus, had been reduced to nine. Now it was reduced even further.

These new raw recruits were novices and relied on the advice and guidance of their more experienced colleagues' as the all-superior German armaments had them at a distinct disadvantage. They were not being helped through the wisdom of the laid back English bureaucracy, who continued to view everyone else but themselves, as cannon fodder.

'What do you think of this, Bill?' Or 'what would you do, Bill?' he was asked.

With nothing to measure experience by, they'd little choice but to ask battle-hardened crews upon returning. There was no shortage of ashen-faced listeners.

Occasionally Bill was challenged, 'I hear you're the expert around here on how to fly a Lanc.'

'I wouldn't say that,' answered Bill coolly.

'You don't look much of an expert to me; they say you break the rules.'

'Do they? Who are they?'

'Yeah, well I think you're wrong. Rules are there for a reason, not to be broken by larrikins. You put the rest of us at risk.'

Bill looked hard at the young man. He was obviously scared shitless and unable to handle what lay in front of him. 'Rules are for stupid people who can't decide for themselves. You do what you're comfortable doing, I'll wage war my way for as long as fate allows me.'

The young man sneered and turned away, as he did so Bill noticed he was in need of a haircut.

'And probably,' said Joe continuing the conversation as if without interruption. 'An over abundance of cheap buttercup yellow paint hasn't helped either.'

Training planes were painted bright canary yellow. No one was sure why.

'To make them stand out and be easily recognisable,' said the powers.

This colourful idea was intended, not so much to enlighten everyone's day, but to highlight how training planes were unarmed and being flown by inexperienced young men with 'L' plates.

This practice was hastily discontinued, with planes repainted with any anonymous colour available, but not before too many had been shot down by enemy fighters.

Bill became thoughtful. 'No doubt, eight-foot Hun troopers and his ally Tojo saw an opportunity too hard to pass up.'

Joe stiffened. 'Shooting kids down to shorten the list of prospective pilots. Bit low.'

'Yes it is. It's also in direct contradiction to the Geneva Convention.'

'They shouldn't have done that, you know,' reinforced Joe. 'Evil, that was.'

Bill sucked on his pipe in agreement. 'I agree wholeheartedly. Tojo has a proven lack of regard for any such niceties as the Geneva Convention. But eight-foot bloody Hun troopers should know better!'

'It goes without saying, but then again, this is war,' said Sandy.

'That you've survived so many ops, Bill, when others haven't,' said Joe. 'Is surely testimony to you being a better pilot.'

'Better, or perhaps luckier.'

'Don't forget us,' said Sandy. 'We're his crew and we might have had something to do with it.'

'Agreed. And as bomb-aimer none could do better. But when Bill makes that split-second decision; veers off to port, or drops us out of the sky, he might be saving our bacon.'

They had another round in reflective silence before it was agreed. If nothing else Ballsy Bill had been raised even higher in status in the eyes of his fellow comrades. Bill remained so bitter his face could have been shaped from granite. He was one of few NCO's flying Lancaster bombers in the British Air-force. Perhaps his short history of being anti-establishment had caught up with him. It certainly hadn't helped his cause for promotion. Doing it his way had been fulfilling. It had probably saved his life and the lives of his crew but had also created enemies. As had telling his superiors in no uncertain terms to go jam it, ram it, stuff it, bugger it or go do it themselves.

'Perhaps they don't feel warm and fuzzy around you,' laughed Joe.

But he was right. Bill was a bit too no-nonsense, straightforward, and far too outspoken for consideration as an appointed leader for His Majesty.

'Perhaps they want more of an establishment man,' offered Sandy.

'Yes. One who stands and salutes at every opportunity, even when on the other end of a telephone,' laughed Joe.

Ben was more serious. 'Aye, someone more prepared to toe their wee line, Skipper.'

'Someone who doesn't mind remaining oblivious to their lack of virtue, and can remain blind to their craven self opportunism you mean,' ranted Bill.

In the air Bill commanded a crew of six others, but on the ground he was required to salute each and every one of them. They'd become officers and outranked him.

'You can be a sonofabitch, Bill. Admit it. It's one hell of an unenthusiastic sloppy salute,' laughed Joe, as he drained his glass and ordered another round. Such was the hierarchy of the RAF either his crew had to invite Bill to their Officers' Mess or they had to slum it down to his level.

Sticking feathers up your bum doesn't make you a chicken but congratulations, you're one step closer to hitting bottom..(Fight Club)

Chapter 38

DRESDEN FEBRUARY 1945

Alice was heavily pregnant with their first child and Bill worried the birth could happen any time. General Eisenhower had announced the main German defence line had been broken in the crossing of the Rhine in France but had added with some caution; I'm not writing off this war—yet.

Further afield in Japanese waters the landing of American troops on the island of Okinawa Sima was yet to be confirmed.

'We might not have to learn German or Japanese after all,' observed Fred.

But not everyone, including Alice and Lily, were too confident of the outcome.

'We might all be pushing up daisies before long, that's for sure,' said Lily, as she raised her hand to cover her smile.

Bill's next op was one that would haunt the RAF, Churchill, the government of the day, Air Marshall Butcher Harris, and even those who took part, until their dying days.

In the NCO Mess, Bill shared drinks with his crew. There was an inflexible division between officers and other ranks in spite of them all carrying out the same jobs and relying on each other for survival.

'If nothing else the powers are consistent,' said Bill. 'I'll give them that.'

More interest than usual was taken in their briefing as it was rumoured to be a doozy, light flak, and minimal night fighters. Where had Bill heard that before? When he looked around the room he noticed the outspoken young man in need of a haircut was no longer with them. He felt sad for the loss of another crew if only it might have been avoided.

The map on the end wall was uncovered. Their target – Dresden! The seventh largest city in Germany was well known for its arts and manufacture of fine porcelain.

Joe was distraught. 'That'd be right,' he said. 'We're going to smash their good china and set fire to their oil paintings.'

There had been a shift in tide of opinion against Butcher Harris along with the cancellation of Bomber Command's campaign medal.

Propaganda meals were scheduled for early evening, as all squadrons were active on this op. First wave crews were due to bomb at 2200 hours.

The target selected as aiming point on the wall was Ostragehege Stadium of the Dresden Sports Club (DSC), located near the main railway bridge.

Bill couldn't help but notice shiftiness of the officers. Implication was one of almost 'foul play.' Feeling among all crews was strong. It was the only time Bill had ever felt the mission was unusual in a distasteful way.

Each squadron of about twenty aircraft were to cover an arc of about thirty degrees. That meant a great deal of co-ordination. Each would fan out from the stadium like spokes of a wheel towards the greater perimeter of the city. Their bombing area was shaped like a wedge of about two square miles but that was Ben's problem, not Bill's. Already Bill had decided if, for whatever reason they couldn't find their target, they would drop their bombs placed as best they could. They certainly wouldn't be going around again, nor circling any pretty markers or bringing bombs home.

Screw their ordinance!

It was supposed to be a five-hour flight with brisk northwest tail winds. Ben looked concerned at complex changes of direction, supposedly to keep German defences guessing.

'Not much the Hun doesn't know about us, Bill,' whispered Joe.

'They'd know more than us, that's for sure.'

There was considerable despondency among crews at the instruction all guns would be emptied facing downwards at their target after their bombs had been dropped.

'Reasons for this attack are murky at best,' said Joe quietly.

'Agreed. We know we aren't targeting a munitions train or factory. It's a bloody sports stadium.'

Tide of public opinion had already begun to swing against Churchill and Bomber Command. Their cronies were seeing many of the despots and incompetents as having too much power.

'Why this unindustrialised city, Joe?' asked Bill.

'To terrorise the Germans, perhaps, or impress the Russians, by showing what Bomber Command is capable of. Who knows for sure? We'll never be told the truth!'

A more factual example of Bill's contempt for 'authority' without 'responsibility' could never be expressed clearer than this impending raid—the firebombing of Dresden.

Even before he took off, Bill felt this op shouldn't take place.

Dresden had never been a military target but history has shown the powers maintained above all else Dresden as suitable. But suitable for what? Retaliation for the Blitz on London?

Bit late for that.

'Perhaps it's a case of an eye for an eye, a tooth for a tooth?' asked Joe.

'That 'Butcher Harris' has a difficult job to do has never satisfied me,' said Bill. 'Two wrongs never make a right. You learn that as kids.'

Bill sat on the runway with a heavy heart waiting for the all clear to take-off.

He knew he would soon lumber down the concrete path, striving for the speed necessary to defy gravity. But he also knew he was to be part of a gigantic thousand bomber raid on a city of sleeping civilians. His cockpit felt colder than usual, even rugged up as he was the chill bit right through him. The metal felt harder and colder to touch than usual and he knew when he achieved his maximum height, the condensation would freeze and make it feel even more unfriendly.

What a miserable way to fight a war.

They were to fly over Dresden in waves over a period of twelve hours with Yanks in Flying Fortresses coming in last. Pathfinders were to arrive first, and their job was to drop aluminium window foil strips in hundreds of thousands. Ben was aghast. 'Aye, the sky will look like it's raining confetti.'

Then the Pathfinders were to ring the city with incendiaries. Red flares dropped from about 3,000 feet were intended to guide in about 200 bomb-aimers. But to ensure the most successful marking possible, an additional 1,000 pound target indicator canisters were also set to burst at approximately 800 feet. They would continue to scatter plumes of light like enormous red Verey flares all across the city.

The marking complete, accurate bombing was to commence from about 18,000 feet to be followed by another wave, and then another, and so on shortly after 2200 hours.

'Tally-ho!' was the signal to start. Co-incidentally a call closely identified with English blood sports.

'Fucking typical,' said Bill. 'The aristocratic elite are to get their jollies at their desks listening to Master Bombers as they imitate the baying of hounds and the rally call to hunt the fox.'

Master Bomber would announce, 'Come in and bomb the red glow!'

'How bloody appropriate is that! Pity it wasn't Hitler's Bunker or U-Boat pens,' muttered Bill.

The first wave crossed Dresden from northwest to southeast, with the wind behind them. The night's weather was better than expected as for once intelligence was correct. No sign of flak or enemy fighters. Bill's scheduled bomb drop was to take place during a segment of thirty minutes between 0115 and 0145 hours. His instructed bombing height was up to 20,000 feet, higher than the first wave. Pathfinders continually dropped more aluminium foil windows.

'Why?' asked Bill, 'there's not a sign of anti-aircraft guns nor Luftwaffe.'

Sandy didn't need co-ordinates. A blind man without a seeing dog could have found the target. The hellhole could be seen glowing on the horizon from about fifty miles out.

Each wave in the convoy dropped its load.

'We're not quite wing-tip to wing-tip, Ben, but that's how it feels,' shouted Bill.

An average of eighteen aircraft a month were being reported lost due to collisions. Heaven only knows the correct number, thought Bill.

With hundreds of Lancs passing over a target in the course of thirty minutes, without navigation lights and at night, he was surprised collisions didn't happen more often.

Six miles from the target other Lancs were clearly visible; their silhouettes reflected as a dark mass through the eerie rosy glow.

'It's almost daylight conditions prevailing, Bill,' said Sandy.

Bill could see Lancs everywhere; to his left, right, below and above. 'That this airspace can accommodate so many Lancs seems impossible. Any closer and our tail gunners can shake hands,' said Bill.

The break in weather not only lasted it improved. Bill crossed Dresden from the southwest, lights from the monstrous fires created by first and second waves of bombing were so visible. Any further marking was superfluous.

'Thirty seconds, Bill,' warned Ben. 'You'll be turning the wee plane five degrees on my next call.' Ben's job was pinpointing his calculations of wind, height, speed and position. For once Bill didn't need Ben. But his immediate concern was to avoid collisions with other Lancs. They were all jockeying to prepare to turn onto their headings for their bombing run. Nervously he kept eying above and below.

'Turn the wee plane now, Bill.'

The bomb doors were opened. They felt a sudden rush of hot air from the cauldron below, which wafted worrying smoke smells in with it. After a few minutes it was up to Sandy, 'Steady, steady—bombs gone!'

The plane lifted its characteristic surge but Bill was pleased the bomb doors could be closed. Their bombs vanished into the maw of fires below. Why bother? It seemed utterly pointless. Like adding a matchstick to an already roaring bonfire!

Then one minute to hold for the photoflash, even that was hardly their usual concern. No flack, no fighters. Nothing! There were only the stars, obscured by smoke.

As Bill passed over the firestorm, intense heat rose from the burning city below. It pummelled his Lanc. Below everything appeared as in a shimmer of a manufactured heat haze.

'My God,' croaked Bill, 'what the fuck are we doing? Civilians are being vaporised down there.'

Over the target Sandy reported it was as if he was looking down the throat of a monster into an active volcano, with masses of molten oranges and reds. Even on the outskirts there were sparks and embers from the fires below flying around them at about 18,000 feet.

It seemed superfluous to drop their bombs but as instructed they had. Then afterwards, instead of returning home as they had on every other occasion, operational instructions after dropping their bombs, was to turn and fly back over the city a second time. They were told to return lower over burning buildings and discharge their guns down into streets and houses below... 'to cause maximum confusion in the civilian evacuation from the east.'

The city had been asleep. Its only occupants were older men, women and children. Burning bright as day, sparks were visible flying in every direction. The whole of Dresden was an inferno.

Sandy was incredulous. 'It's the perfect firestorm,' he said. 'Entire chunks of red-hot matter. Everything is ablaze. Below us streets of the city appear as a latticework of fire.'

Bill was overawed by the spectacle. The confusion and panic below would have been unspeakable. He visualised people, their guts spilled onto pavements, the half-dead draped over steaming antique fountains.

Aircrew, as high as 20,000 feet, felt the effect of the pillar of heat produced by such a blaze. Suddenly, Bill felt unworthy. What if circumstances had been reversed? He knew of no instance when Jerry had bombed London and tried to machine-gun his parents in Barlby Road.

He watched a Lanc go down. It exploded but there was no flak. Bombs from a Lanc above must have hit it.

Bloody fuck! There'd be little point in baling out into that hellhole.

'Everyone keep your eyes peeled,' ordered Bill, 'Concentrate or it could be us next.'

Bill normally flew lower than instructed but not on this occasion. If he broke the rules it was to maintain optimum height or higher. On his return run he didn't fly as low as instructed.

Bazza didn't want to discharge his guns down into civilians. 'Only if you tell me I have to, Bill,' said Bazza. 'You'll need to give the order, mate.'

Bill gave the order. It was war! We all seem to be walking in circles to our destinations, he thought.

As a result of that order, Bill personally discharged the lower front guns, which were normally unmanned. It was that, or order Joe or Sandy to do it. Both Joe and Sandy were unwilling to volunteer.

In his defence Bill was twenty-one years old, newly married with a baby on the way. He didn't want to be there. They flew back over the burning city, what remained of it, but at height. It was the only time he'd ever been ordered to fly low and hadn't. It was also the only time he ever returned over a target. The scene below was one of utter devastation. Rarely did aircrew see results of carnage caused. Dresden was an absolute fireball, followed by a continual drenching by emptying each plane's .303 calibre Browning machine guns into streets below.

The pandemonium of burning homes and innocent people dying in the most awful circumstances, etched into Bill's brain, never to be forgotten. But it had been an extraordinary instruction.

Bill forced the big guns downwards. His thumbs hovered over the fire buttons. For a moment he delayed. He knew at this height he wouldn't see people on the ground but certainly he'd no intention of trying to hit them. At a time when all he could see was collapsed structures and flames he held the triggers down.

Yes, he followed orders, yes, they discharged their guns, but Bill had no intention of deliberately machine-gunning civilians. Vibration and noise of the discharging guns echoed up his arms and throughout his body. But all he could think of was some unfortunate grandparent or child who might be copping it below.

Fuck this! I didn't train to be a pilot for this shit!

The warrior's spirit in him spent, he wept as he carried out His Majesty's orders against humanity. His heart pounded in his ears. The guns leapt and jumped in his hands for what seemed much longer than thirty seconds.

His guns were quiet. The awful deed was done. When he took his seat again in the cockpit he felt unclean. His distaste for the powers had reached an all time high. He came close to a pinnacle of extreme disrespect. His rebellious nature, stemmed from his origins in the slums of Notting Hill resurfaced, in a burning desire for fairness and justice.

Bill had gone full cycle. Disparity between optimism of an eager young man who had set out to be a pilot, and bitterness of the man he'd become, couldn't have been greater. Originally he'd longed to better himself and fly for His Majesty, wanted to become an officer, but now he wanted to be discharged. Even being put in a holding cell looked good to Bill. Anything to escape this madness.

Waging war from the air is often seen as random and indiscriminate even when it isn't but Bill saw Dresden with the clarity of cold blood and he suffered a crisis of conscience.

Their return flight was without incident. As aircrew were debriefed there were none of the usual callous remarks. Most kept their heads down. Many felt bitter but used to their usual horrendous losses over Germany celebrated a mission that for once had been almost a bloodless coup.

In his flying 'logbook' under the section headed 'TARGET' Bill wrote the words; 'women and children.' * 11th Footnote

This caused an awful fuss but Bill didn't care. He was threatened with every manner of nastiness ranging from suspension to court martial if he didn't delete those words from his logbook. Many including Bill were of the opinion Air Chief Marshal Sir (Butcher) Arthur Harris had committed a deliberate war crime. He deserved the title of 'war criminal' and yes the bombing of Dresden did happen and was a 'war crime.'

'Nice,' Lily said. 'Rub out the words to remove the problem, Son.'

Bill stuck by his principles and refused to change his logbook entry. He rebelled against the system in the only way he knew how. In his heart he knew what they'd instructed him to do was wrong. He also knew he was under directions to carry out orders.

'But at what stage does that become an option, Alice?' asked Bill. To his question no one had an answer. In later years 'conscientious objectors' would be publicised but always would be the inference of cowardice.

Bill's perspective of the RAF had further changed to bitterness and darkness. He had wanted to belong and become a pilot but now saw pettiness and stupidity. Buffoons high up with no guts were committed to send young men to die, often in a plane that through lack of maintenance should have been grounded.

'Too many snouts in the trough, without courage and with limited experience,' asserted Bill. He began to distance himself from what he considered to be travesty in the loss of humanity.

Fate had another surprise in store. After his logbook debacle came criticism about the quality of Bill's pictures. Over exposure due to the extreme conditions over their target, even though on that occasion he'd gone in at their recommended height. Bill couldn't believe what he was hearing. But he no longer cared. His mind was elsewhere.

The officer was angry. He shuffled his papers and glared at Bill.

'We were there. We were hardly anywhere else,' snarled Bill.

He fixed the officer with an unseeing thousand-yard stare. As if it was Bill against all of them but he felt dirty, tired and worn out.

They looked fresh, spruced and smug in their safe roles. Bright eyed and with pencils poised, they faced him and other weary warriors, about to interpret for the powers their own politicised versions of a night's carnage.

'This de-briefing is over for now, Flight. You are dismissed.'

Bill wondered if the officer shouldn't hang-a-cup-out-of-his-arse to catch any excess, but nothing would fill the hole he felt with the loss of Olly. Horror and dismay drove through Bill in a freezing wave.

Bill was instructed by the powers to change his logbook entry. He refused.

'I have carried out my orders correctly and as they were given to me—Sir, but let there be no misunderstanding on anyone's part—Sir.

The 'TARGET' in this instance was non-combatant civilians, and not a military installation. The logbook entry stands as correct—and stays—Sir.'

No one could ever claim what happened at Dresden was an accident. Bill felt ashamed he'd levelled himself to the Krauts. He'd bombed women, children, aged, dispossessed and homeless. The deed turned the op and Bomber Command into a byword for slaughter.

There was no doubt in Bill's mind, if the Allies lost the war, Dresden would have been declared a war crime and Sir Arthur (Butcher) Harris shot or hanged on a gallows for his part in it. *14th Footnote

*****

How many died? Who knows the number?

On your wounds we see the suffering

Of the nameless ones who burned here

In a fiery hell of human making...

This plaque commemorates the wording from the Dresden Memorial (2002)

*****

I, and the public know, what all school children learn, those to whom evil is done, do evil in return... W. H. Auden

Chapter 39

BILL'S CHAUVINISM MARCH 1945

The war was nearing its end but there were more ops. There was frivolity in the air among civilians but Bill felt down. He'd become an automaton. Because the Allies felt more confident orders were to bomb more frequently and in daylight. Bill felt crews were being wasted dying on unnecessary ops. He was of the opinion the powers had accelerated getting rid of ordnance.

Bill had taken Alice to RAF parties inside Base, where there was warmth and conviviality. More importantly there was plenty of good quality free food and alcohol, which otherwise was difficult and expensive to attain even through the black market.

He stopped taking her, with fresh distaste, when he realised she might be offended. Not so much by drinking, chaotic dancing, laughing, throwing up and falling down but by the insipid looking balloons, which were not balloons at all. They were in fact Durex condoms, although he doubted Alice ever noticed and he never asked. Their bedroom events took place in total darkness and beneath covers. Alice had felt her husband on many occasions but she'd never seen him.

Early in March, Bill's crew bombed the Refinery at Harburg, a satellite city of Hamburg. There were no German fighters sighted but flak was moderate. In spite of everything he vomited from the stress. Lancs were being lost in droves.

'What's wrong, Bill?' asked Alice. 'A penny for your thoughts.' He paused, as if to let her know her own thoughts were as precious to her.

'The certainty, Alice, is people are dying and for all the wrong reasons. One of the other crew's navigators has announced he won't be flying anymore.'

'Can he do that, Bill? I would have thought he'd be stamped LMF.'

'Apparently he can. He contracted VD and someone said that's an instant disqualification from flying. I couldn't help but think about Olly. Perhaps they botched his Canadian paperwork and Operations wasn't aware. If Olly was alive he might have been able to get out.'

'Oh, Bill. That's dreadfully sad. So many people have lost their lives. Cut short in their prime. Poor Olly.'

'I think of him whenever I look at the thermometer he gave me, but the memory is too painful.'

Alice gave Bill a big hug.

Bill never asked what happened to the thermometer. When he looked at the scullery wall the thermometer had been taken down. Strangely, after that his memory of Olly slipped, along with Canada although it remained a distant beacon in his mind. He felt guilty he'd allowed the memory of Olly to fade like that.

March 1945 was a busy month for Lily and Fred at home in Barlby Road.

Wednesday 28 was particularly unforgettable and demanding for Alice. For her it hadn't started out well. The tiny flat had been an oasis in the middle of a formidable battleground for years. Now Alice was about to give birth in the little room. There was no doctor. Lily helped as midwife. And Fred was set to boil water and collect up clean towels as fast as he could.

Not that they needed to get ahead of themselves, yet. This was to be the main event and would drag on for some considerable time.

'Not directing blame here,' said Lily, tired from beneath heavy eyelids. 'But I think your baby might want to spend his twenty-first birthday in there with you, love.'

'Talking of birthdays perhaps that's why mothers never forget their children's?' joked Fred. 'All that reminiscing about pain and suffering they've endured, remembered, year after year.'

'Perhaps that's why God gives women children to bear, you mean,' quipped Alice between grimaces. 'So that later death doesn't come as such a disappointment.'

'Look upon it more as God's way of punishing women for having sex,' sniffed Lily.

Bill knew the birth was about to happen but he was away. On daring daytime raids he dropped his bombs on U-boat pens, an oil storage depot, two benzol plants near Vegesack, and the railway centre of Paderborn. He was particularly pleased with the powers' decision to bomb the U-boat pens.

'We should have made a concerted effort long before this,' said Bill. 'We might have saved a lot of ships and sailors' lives if we had.'

Daylight ops had distinct disadvantages. It was easier for the enemy to see him and he watched other Lancs go down, when at night the dangers were there, but less visible.

There had been a few modifications and improvements made. Bazza had an increased chance of survival because of the introduction of a parachute worn as part of his gunnery seat. No need to exit now via the fuselage after he'd retrieved his chute from the hook on the wall. But for the RAF there was a drawback. Some rear gunners with survival on their minds flew with their rear turrets swung all the way around should they need to bale out.

Bill had always been a male chauvinist, and that was all right with Alice. She enjoyed her status quo that way. Alice was a woman who liked a man to lean on, which was not unusual among the women of 1945. It would be another twenty years before some burned their bras. In Bill's day the honourable male attitude was to encourage the finer sex to be wholesome, clean and to nurture children. Also they should cook and keep house for their man. This suited Alice because it was the way she'd been raised.

When Bill and Alice went out for a stroll she took his left arm and he positioned himself kerbside. Alice laughed. 'Keeping your sword arm free, Bill.'

'Yes, and I'm protecting you from mud thrown up by horses and carriages of yesteryear.'

Bill gave what he considered to be basic respect to women. He stood up in their presence if they entered a room, or joined him at the table. He gave up his seat on public transport, held doors open, and spoke politely after first removing his pipe from his mouth. He used miss or ma'am to address them and while he undoubtedly had a 'potty mouth,' he rarely swore around women. He abhorred the sight of any woman who smoked or became intoxicated in public. Bill felt these were not suitable feminine pursuits.

There were many men like Bill, including his dad Fred, Uncle Charlie and others such as his colleagues on Lancaster crews. But shades of grey surfaced when a female officer was on Base. Bill was expected to salute her.

'No way. I don't give a toss if she's related to the Queen or she's Sir Winston Churchill's mistress. No way!'

True to his word, Bill never did salute a woman.

This posed another problem for the powers, but Bill was adamant.

'Might be a bit unfair on female officers, Son,' Fred pointed out.

'I'm not the only member of flight crew not to salute women,' Bill said defensively.

Coming face to face with a female officer, he'd smile, take his pipe from his mouth, and address her as miss, or ma'am, but no salute. Bill was not alone with his views. There were plenty of outspoken young men like him. Under no circumstances was Bill prepared to salute a woman and to some extent he got away with it.

Few saluted on operational squadrons anyway but had an officer complained it was likely Bill would have faced a charge. Since he wasn't up for a commission he didn't care much. There was one occasion when he was brought to account and asked by an officer why he'd failed to salute an adjutant. In all honesty he'd never seen her.

'Because, Sir, the morning sun was shining full into my eyes,' answered Bill.

Why tell you the truth? You're not worth it.

There was almost a flicker of a smile before, 'Admonishment.'

An admonishment meant nothing.

An Irish colleague was a fellow pilot in the same squadron. Patrick was a character. In the Mess he sang along with one song. He insisted on teaching the words to everyone.

'I'll take you home again, Kathleen' was a rousing favourite with most crews. As an Irish ballad it was without equal and when Pat sang it, it was as if he was home in the Emerald Isle.

When they knew Pat better he proudly showed them a letter he carried from the German Luftwaffe command. They politely declined his application to join them on the grounds they wouldn't allow an Irish born affiliate to join their Luftwaffe. With a few ales under his belt Pat was quick to add his thoughts to any who cared to listen.

'See! The Germans. Now they've got fuckin' principles. Only you English would take me!' screamed Patrick.

Looking for a deeper meaning of motive Pat might have seen that letter as his insurance policy. If the war was eventually lost to the Germans, which at times looked likely, that letter could become useful to him.

'Not a bad ace to have up his sleeve if he was shot down over Germany either,' as Mack was quick to point out.

On the ground Bill needed to get about and this was best achieved in the form of mobility. His first 'roller' was an aged model T Ford. He paid £2 for it. Hardly a high- powered projectile at nought to thirty miles an hour in a staggering sixty seconds. As distinctive as the new car smell it hadn't, it reeked of cigarette ash, stale booze, sweat and unwashed bodies. Innards of seats burst forth in all the wrong places.

Petrol was rationed. As a result even horse-drawn vehicles were back in service. Good news for farmers as the price of a horse had risen from £40 to £60. Bill and his crew were unmoved. They preferred four wheels and shamefacedly stole high-octane petrol from forty-four gallon drums on Base to run his car.

It interested Alice how Bill's moral scruples dissolved in the face of this attraction.

He was content to accuse others of corruption but stole petrol. Dyed a different colour, it was green to distinguish it from petrol sold at garage bowsers, which was sort of red-ish. The car rocked crazily along narrow lanes but ferried him and his drinking buddies wherever any grand little pub was open.

Someone thumped the piano with dynamic energy. The old Joanna they called it, as they bawled out any chorus to anything everyone seemed to know.

One evening, pissed and grinding his gears unmercifully, his nose almost touching the windscreen, Bill hit a cow. He wasn't driving fast. Only travelling a little faster than walking pace but the cow stood right in the middle of the road, and he hit it, hard.

'Perhaps he'd had a memory lapse? Thought his car was a Lancaster, instead of braking he should drive on through?' laughed Fred.

Bill was amazed how the cow mooed and limped away unhurt, but left him with a wrecked car. What issued from his lips sounded like Swahili but he and his mates towed the car back to Base for some customised, cheap modifications?

The 'roller' was never quite the same after its close encounter with the unyielding bovine. Bits kept dropping off and he added drawing pins as a feature to hold the roof up on the inside of the cabin. Whenever he went over a bump, drawing pins popped out and the felt lining came down over his head and obscured his vision. Most times he would drive his beloved car with one hand on the wheel, other suspended above his head to hold the roof up. He needed a third hand for gear changes.

So there was Bill, three sheets to the wind and grinding his gears. His stocks of protruding jerry cans over brimming with illicit high-octane petrol headed to the pub or go see Alice and his newborn baby boy. His flight crew promoted the concept as to how it might be a good idea for him to obtain a driving licence for the car. These were issued from local police stations.

The police sergeant in charge took the applicant out for a bit of a test drive to check how he handled important mechanical bits, clutch, brake and any hand signals individuals might feel were important. There was no highway code as such, it being left to good old basic common sense. Irrespective of his abilities in the cockpit, Bill procrastinated and kept putting off the application for his driving licence. It became more than a small chore.

Maybe it was his reluctance to be tested by officialdom or perhaps a fear of failure and ridicule, as when he'd experienced difficulties turning his Tiger Moth on the ground.

'Go on, Bill. Be a dare devil,' laughed Joe.

'Aye, a couple of Scotches might improve your gear changes,' suggested Ben.

One day, a little irritated and running late from de-briefing he presented himself to their local police sergeant in full flying kit, with a polite request to be tested. He'd picked a good day for it with clouds chasing in from the coast. Amid mud and drenching rain he splashed up to the door. The light was already failing.

It was too much of an imposition for the older sergeant who was greatly intimidated by this young man from the flying machines piloting Lancaster bombers. The Sergeant wrote out a driving licence on the spot. Not a single test question was ever asked.

Bill was in his element and wanted to party. He felt he'd something to celebrate.

In the Mess in front of the welcoming fire he kicked off his wet outer clothes; vapours rose from about him. Edging into comfortable warmth, he became quickly inebriated.

Sometimes intoxicated young combatants did silly things that seemed quite reasonable at the time. Not unlike Olly's entry astride his motorbike. Like party animals off the leash they decided to go knock on death's door. This they did to check out what a man with the name Mr Death looked like. He was a reasonable man who didn't object to their sarcasm and silliness as he respected their uniforms. He asked them to come inside where he explained without malice.

'My name is spelt Death, but my family pronounce it, De-ath.'

Bored, Bill's mind was already home in bed with Alice.

Love is the delusion that one man or woman differs from another...(Henry Louis Mencken)

Chapter 40

VE DAY AND VJ DAY DECLARED

As with all good things, everything comes to an end. President Roosevelt had died of a cerebral haemorrhage in April, which gave Harry S. Truman a crack at the top job in time to bomb Hiroshima in August.

In London a summer shower of rain had fallen. Bill felt down, as he knew it would be blanketing the countryside in a misty vapour. He longed to be out of the city and close to fields where he could smell the musty wet grass.

In Berlin, at the end of April a sad and lonely little man with an enormous ego, committed suicide. A gun in one hand, a cyanide capsule in the other.

For the lucky ones, the horror paled towards year's end. For others one day walking and talking, next day cold as stones. The peace everyone had yearned for arrived.

'A bloody shame,' Fred said. 'For about seventy million people, who aren't fortunate to be around?'

At the end of history's greatest conflict and monstrous absurdity, people grappled with the horrors while benighted fools counted their new found wealth. City buildings throughout Europe had gone from symbols of culture steeped in rich history to vast shattered rubble. Britain's backbone had crumbled and Germany was a smouldering ruin. Ordinary people who lay dead on the field of battle had lost their names while rich men had grown richer.

Reports of men believed dead, who had actually survived had always been rare. Now they'd become even more rare. Bill was incensed. 'Death should command respect even among those of little value.'

With the world largely at peace church bells began to peal for the first time in seven years. As the wireless relayed continuous victory celebrations between Vera Lyn's songs, there were tears in many eyes and a lump in as many throats.

It dawned on Bill and Alice they could contemplate a future beyond the war.

Somehow to Bill it seemed hollow. He'd come to realise the RAF's guiding principles that had been his deliverance had also ironically, been his undoing.

Alice grasped on quickly but Bill had difficulties. He was unable to clear his mind of all the young men he'd known who had died along the way. The images of fresh faced youthful males who weren't there anymore kept flashing into his brain. The speed with which orderlies removed personal possessions and prepared for the next intake, unsettled Bill now, more than before.

They certainly didn't need time out to reflect, thought Bill.

Empty beds on any morning were testimony to airmen who had not returned and it took Bill a while to adjust. Too many times he'd enquired of someone only to be told, 'Sorry, Flight. They didn't make it back.'

Lockers contained heartrending proof of their short lives such as a book with its marker at a favoured page, or at a point when he tried to lie down and sleep. Well-thumbed photographs of loved ones, even a hasty letter written should that last battle with the Luftwaffe be lost. Then the well-oiled process kicked in, personal effects posted the day after the Air Ministry telegram had been sent labelled 'Priority.'

Bill's farewell letter, the one he'd left in his locker had been vexing for him to write. Now as he stood there and held that epistle in his hand, he realised how lucky he was to be one of a rare few in a position to destroy their note. It had taken him a long time to write it, but that had been good because it helped consume difficult to sleep hours. As Bill held the letter close to him his ghosts gathered. He felt cold, his position irretrievable. Faces of lost comrades like Richard, Tache and Olly became blurred.

'I felt embarrassed and a bit like an officer who'd let a bunch of Lancs go missing,' Bill said to Alice.

When he was able to put aside some ghosts of his war, he realised he had a secret desire—to lead a parade.

Bill was granted his big opportunity. Joy mingled with relief.

Everyone partied and celebrated for days. In sombre moments they thought about those they'd known who hadn't made it through the conflict.

Women shed tears for lost loves. Crews hid their faces from prying eyes as they broke down and thought of comrades who hadn't survived. Some insisted on pouring a drink in the Mess, a lone glass to stand as testimony to their fallen comrade, as the cadets had done for Richard in Canada. Emotion poured from their singsongs.

It was to be Bill's privilege to lead the squadron in a public march prior to being de-mobbed. The Victory Parade was due to commence early in the morning.

The night before Bill tied a good one on with his mates and crew. He felt the loss of Olly and drank more than his fair share, unable to resist the desire to crawl into alcohol's womb.

Come morning, as a result, he was a bit fuzzy and not firing on as many cylinders as he would have liked. That Bill had a hangover would fail to even hint at the quaking nausea and his black despairing outlook.

'Oh, Bill,' said Alice, 'you're going to be late, so late.'

'He's not the only one feeling his age,' grumbled Fred. 'We'll all be feeling old before this bloody spectacle of a pageant is played out.'

Bill was rat faced. His hangover crept up the back of his neck. He would have preferred to not lead the parade, instead kick back and relax in the foetal position, unshaven and unwashed. But he was committed to lead that now not so desirable march.

To make matters worse, he had to hurry. He suffered inside the collar of a shirt that already choked him. He barely made it to his place in front on the parade ground.

As he looked at the disorderly line of men behind him he realised this might well be his last parade. At boot camp this wouldn't have been classed as a parade at all.

Boots no longer gleamed, buttons no longer shone like precious metal, and not every man stood stiffly.

Momentarily Bill missed the rigid discipline. He'd moaned, as loudly as anyone at the bullshit. The scrubbing and polishing, marching and drilling, but with it gone, it seemed more meaningful than before.

The officer inspected the men before him. He was a younger man with a kind, but hard-boiled face. He stood in front of Bill dressed in his best blue. It was almost as though he hesitated, as if he was unsure whether to stop or not. He glanced up from under his eyebrows.

What he saw was a hastily selected dress uniform under a peaked hat. And he was not greatly impressed. He reached forward with his baton, gently pried the peak of the cap skyward to reveal the true glory of what lay beneath. Bill was pale and sweating, his red rimmed swimmy black eyes, like piss holes in the snow, were set in an unshaven face framed in mottled saggy grey skin. The whole swayed gently before him, which cast some doubt on the chance of Bill staying upright. He could see the pulse gallop at Bill's temples.

'Oh—yuck!' It was his only comment.

Bill spent a lot of time waiting around. Airmen everywhere lounged, loitered and chatted among themselves. Some occasionally took a stealthy drag at a cigarette,

Bill puffed his pipe. Sergeants with short memories and uncontrollable urges called names from lists. Instructions were given as the great filter tank of the RAF sorted itself out. Some were for re-muster but most for discharge from service.

There was a token discipline as men put in time. They waited, waited, waited...

Briefly Bill considered a career as a civilian pilot.

Most of his crew were unsure what to attempt. Of them all, the only man who had made up his mind was Frank. He'd decided to become a policeman.

'That'd be right,' said Joe. 'He's the biggest tealeaf of the group. Can't even stop himself from taking another man's clean shirt. He'd steal stink off shit, and what's he going to do?'

'Become a bloody copper,' echoed Bill, as he diced and ate an apple. The policeman who had stood beside the man selling winkles sprang to Bill's mind.

There was a lot of competition for what few pilot jobs were available and many of Bill's peers, unlike him, had achieved officer status. Some had flown more hours but he pursued the idea until he came up against one insurmountable obstacle.

The Royal Air Force announced he was unfit to fly. They stated he had severe sinusitis of his mucous membranes, a condition considered unsuitable for a pilot.

Treatment comprised regular scraping out of nasal passages to relieve pressure, but this wasn't considered a permanent solution. Severe headaches or even a blackout may result, he was told.

'The bloody goal posts have been moved again,' sighed Bill. He felt no more useful than a used bus ticket.

Many had voiced during war how the powers counted your eyes when it suited them. With the war over there was abundance in supply over demand for pilots.

The process of selection had changed and become more discriminating. Not only did they check Bill's eyes, but also they wanted better than 20/20 vision.

Bill was peeved. He had wonderful eyesight. Two good pale blue eyes with 20/20 vision in both but he couldn't beat this sinusitis of his mucous membranes condition.

The same doctors who had passed him fit to fly in wartime, now shook their heads. To Bill the same organisation that had trained him to fly, expected him to pilot anything in any condition, safe or not, had turned against him.

'Sometimes in such Old Testament weather conditions that even feathered birds, through good instinct wouldn't fly, Alice. But I was expected to fly.'

But now they rejected him. Bill and Alice were devastated.

Often the wind had been strong enough to blow a dog from its chain but he'd been expected to take off with that big bird and his overload of bombs. Regardless of whether he'd a runny nose or not, whether he could breathe through his nostrils or not, he was instructed to fly and no-one had ever suggested or cared he might blackout.

Now any ambitions of a flying career were over and in what to Bill seemed such an unfair and hypocritical way.

It had helped to be good at the job. Bill believed he was a good pilot, but even if he'd been a positive genius in every respect, humiliation and ridicule seemed to lurk around every corner.

Lily sighed. 'You're well balanced, Son. You've got big chips on both shoulders.'

To Bill they'd become a lot heavier. Irrespective the powers were right.

Bill never flew again as a pilot. They had seen to that. * 12th Footnote

'Not only is there no God, but try finding a plumber on Sunday,' Woody Allen

Chapter 41

PARTHIAN SHOT 1946

Trials of Japanese war criminals began in Tokyo at about the same time as Bill's honourable discharge from the Royal Air Force complete with wings and ceremonial sword. When the Air Ministry pledged to send him his medals he told them not to bother. He didn't want them.

'They're only campaign medals, same as those given to everybody else. If I had defeated the Luftwaffe single handed they might have given me something worthwhile, but posthumously.' * 13th Footnote

Bill changed his surname by deed poll from 'Honey' to 'Hickman,' his mother's maiden name. His official reason for the change was to stop harassment and name calling debacles he'd suffered as a child.

'Being called sweeter than sweet got me down. I had an opportunity to do something about it, so I did.'

It was also the surname of Lily's well-to-do brother Charlie, who had given Bill the cricket bat as a Christmas gift. A man he'd admired as mentor and close family confidant.

Bill's dislike for his dad had become palpable and there were other deep-rooted issues. Bill blamed Fred and claimed he lacked substance. He constantly reminded Fred that without his push he would never have left Notting Hill.

'I'm not the bloody dreamer, Dad. You are! You're only good for telling kid's stories!' yelled Bill.

Fred resented his son's verbal onslaughts, but loved him dearly and admired him for what he'd achieved. Fred stayed quiet and never spoke up in his own defence. And Lily disappeared quietly to shed her tears.

Alice suspected Bill changed his name to achieve a mental separation from his dad. But if he did, he never admitted it. To any suggestion twenty-three years might have softened the impact, Bill replied, 'I know how to hold a grudge.'

Fred could never understand why his son was so against him. He was doing his bit as a guard on the London Underground railway. At least it was a step up from dockers in flat caps. In the country's time of greatest need he served the nation as best he could.

He helped oil the wheels of commerce or at least ensured they didn't flounder completely to a halt. And he'd served in the Home Guard. More important to Lily, he was the man. Her-man! He waved a green flag as the all-clear signal to the train driver, and a red flag when it wasn't. He also blew his whistle loud to attract the attention of the train driver to him as he waved his flag.

That other men more worthy than him, with more brains and schooling drove trains and had motor cars instead of bicycles never worried Fred. Lily and Fred were happy. And that seemed to go right over Bill's head.

His 'change of name affair' for whatever reason gave Bill some sort of psychological separation from his dad, or at least closure of some issues. And Lily could never understand why the fuss. Fred had worked hard all his life. He and Lily never had a bank account and lived a cash existence. Their bills were paid, as they were due from little jars and tins on the mantelpiece; the 'rent' jar, the 'insurance' jar, the 'food' jar, the 'electric' jar.

Those jars drove Bill mad. Small differences became bigger ones. Unspoken resentments chilled the air.

Lily and Fred believed those jars provided a style of comfort, a kind of caution for those who had known hard times and feared their return. They expected little or nothing from life and were suitably rewarded. But they'd learned the secret of true contentment; not getting what you want, so much as appreciating what you have. They felt they'd everything; friends, family, love and a healthy grandson.

Bill wanted more. 'Some people shit and get rich. Others are plain lucky, Alice.'

'You aren't wealthy but you have survived the Second World War, Bill. You should think yourself lucky in that respect.'

We owe respect to the living. To the dead we owe only the truth... (Voltaire)

Chapter 42

BACK TO BASICS

His wait over, Bill was de-mobbed. He left the service at the same time as all the other draftees, which contradicted the original government promises. They had failed to keep their pledge and admitted they'd botched the completion of their discharge paperwork.

'They couldn't even get that right,' muttered Bill. 'Although it's probably one of the rare times they've ever told the truth.'

Bill became one of thousands of typically discharged grubby airmen.

The RAF removed his distinctive blue uniforms and fitted him with an uncomfortable demob suit on 27 November 1946 to commemorate the end of the long dark night of war torn Europe. It was a poorly cut charcoal grey garment of stiff serge, which made him feel even less worthy. *What bleeding heart thought that was a good idea?*

Bill remained truly Alice's 'Reluctant Hero.' He had a month's sick leave and was exhausted mentally and physically with sixty-four days holidays accrued. The experience had changed him. Stretched out on his familiar comfy bed in the little room, his cricket bat against the wall in the corner, Bill broke down but couldn't understand why.

'I feel like a piece of stretched elastic and pulled to the limit, it's parted, Mum.'

'Few things are quite as depressing as the process of healing, Son. I'll make you a nice cuppa.'

He cursed his nervous system that was blown to pieces. He'd reached a pinnacle of his life before the age of twenty-three. Why had he been so lucky to survive, physically intact with all his limbs in place and no terrible burns nor disfigurement, when others had not?

Too late he realised he'd called his bosses useless bastards and that hadn't helped much, either. Had he learned to keep his mouth shut and not openly criticise his masters, he might have been promoted.

'Officers received a better quality demob suit, Son,' reminded Fred.

The following week, as Alice dusted and tidied, she noticed Bill's cricket bat in the corner of their room. 'Since you don't play cricket, Bill, should we put this bat away for little John when he gets older?' she asked.

Bill thought for a moment before he answered, 'No, Alice. Throw it out and we'll buy him a brand new one when he's old enough. There'll be nothing but the best for our boy. He won't grow up in the slums like we did. He'll go to the best private schools.'

'Bill, that's wonderful, but very expensive.'

'Yes. He'll also read books. Many books. One day he might become an author, but we'll start him off with his own library; the works of Lewis Carroll.'

Bill knew somehow he needed to move on. There were two major considerations now. He had Alice and they had a family together, a son John William. With them to support, Bill took stock.

A lot had happened. He'd grown up, become bitter after five years, and twisted after ten. But he realised his circumstances had not changed much. He'd owned a suit of clothes when he enlisted and now he had another. He had little or no money, yet money was his yardstick. He lived in a room in a rented flat but had learned to fly, and he'd survived that dreadful war when many had not. He knew he'd nerves of steel. He'd proved that. Now he might need a different style of nerves—to be a parent.

He looked lovingly at Alice as she cradled their son in her arms. 'I'm sure there are great days ahead of us, Alice.'

She gently squeezed his hand, 'I'm sure you're right, Bill.'

'If you haven't thrown away that thermometer Olly gave me, would you mind if we put it on the wall again?'

Without fuss the thermometer was returned to its previous position.

'Whenever I look at it I think of Olly,' explained Bill proudly. 'He was such a fine young man.'

'They all were, Bill,' said Alice, with tears in her eyes.

'They were the best of the best, and better pilots than me.'

'Why do you say that, Bill? You had to be good to survive.'

'No, I didn't, Alice. The good pilots died. They flew by the book and were brought down by Jerry. I was a lazy pilot, I never flew according to the book and I broke their rules big time.'

'What difference does that make?'

'I was unpredictable, Alice. They weren't. Jerry found them when he didn't find me.'

'Oh, Bill.'

'Our world is a sordid and corrupt place and I think it corrupts me. But it's our world. And unfortunately it's the only one we know.'

'Oh, Bill. How can I help you reignite the spark that's been missing so long from your life?'

'It can only be revived with a new challenge, Alice. This is our chance for a fresh start. It's time for bigger and brighter things, an opportunity to escape the bottom-most rung of England's social ladder. I plan to achieve by becoming wealthy.'

'That's wonderful, Bill. But how will you do it?'

'I'm fed up with taking orders from fuck-wits in the RAF. My training has taught me other skills besides how to march and fly. I learned to evaluate difficulties in crisis situations and make split second decisions, when my life and lives of my crew, were at stake. As a boy Lily told me, 'You shouldn't be getting ideas above your station in life.' But now my station in life has changed. If nothing else I'm an ex-pilot of Lancaster bombers. No longer can I be accused of 'wasting time with pipe-dreams.'

Bill felt a sudden surge of confidence. He had new dragons to slay.

'Alice, rather than simply finding a job, and they're queuing for those in their droves. I have a much better idea. What if I become my own boss?'

'That sounds good, Bill. Tell me more.'

'Before the war I was apprenticed to a manufacturing jeweller. I worked with gold and set diamonds in rings. I was good at it, except the day I almost died.'

'Oh, Bill. You must take care, but the diamond business sounds so exciting. You could make me a beautiful ring and I'd wear it always. Together we'll rise above our humble origins and make something worthwhile of our lives! Oh, yes, Bill, please!'

Sudden optimism infiltrated every nerve in Bill's body. This would be a new chapter in their lives. He would be successful for the sake of his family.

That night sleep eluded Bill as he tossed and turned, his thoughts jumbled and confused. Then about midnight he was suddenly jerked fully awake to the loud sound of an aeroplane approaching. As he lay there he tried to estimate its height. Listening, he stared into the darkness and wondered why?

At first he'd almost panicked, his whole body tense as he waited for sounds of wailing sirens and crashes of exploding bombs. But nothing happened.

He snuggled against Alice and remembered when he'd flown over roofs of sleeping English villages at less than 2,000 feet. His thoughts then were of people in their beds below him. Smiling to himself, he wondered if the pilot overhead might be having similar thoughts. And when it passed directly overhead and their roof tiles rattled it amused him.

'Are you all right, Bill?'

The mere sound of Alice's voice immediately comforted Bill. 'I'm fine. You go back to sleep, Sweetheart.'

In the cot their baby son, John William stirred. Soon it would be time for his early feed. Bill hoped he wouldn't become a boring child.

The peacetime aeroplane droned away in to the silence of the night. With a sigh Bill turned over. He closed his eyes tightly against his previous thoughts of war, the canons and horror sounds of his enemy. Now the mighty Luftwaffe was asleep. No longer were they coming to bomb his home and family.

All men could sing and he, Bill, might attempt to whistle a fair to middling tune. He felt confident now. No one should think ill of him if he tried.

Hell is empty and all the devils are here...(William Shakespeare)

Epilogue

Six weeks after the end of the war there was a General Election. Among the rubble of war torn Britain on 5 July 1945 Clement Attlee leader of the Labour Party became Prime Minister. Churchill lost by a resounding majority of 146 seats.

* * * * *

Whenever Bill was asked why he'd flown aeroplanes he answered, 'for nineteen shillings and sixpence a day.'

If asked why he flew Lancaster bombers he replied, 'because they wouldn't give me a fast one.'

Footnotes

* 1st Uncle Charlie managed to leave the bookie business before he lost everything. He settled permanently in a leafy London suburb with his family. He kept his house and car and became employed as a factory supervisor and remained a staunch supporter and mentor of Bill until his death.

* 2nd Rudolf Hess was an expert pilot. On 10 May 1941 he flew a Messerschmitt ME-110 fitted with long-range fuel tanks for the 900-mile flight of five hours. He navigated to within thirty miles of the Duke of Hamilton's residence in Scotland before baling out. After parachuting safely to the ground he encountered a Scottish farmer and told him in perfect English, 'I have an important message for the Duke of Hamilton from my Fuhrer. His plan was to approach the British government through the Duke whom he knew personally. Despite his attempts to negotiate a peace settlement with the British government, he was ridiculed until Churchill ordered Hess to be imprisoned for the duration of the war in solitary confinement. If the British government had handled his intervention differently, might not his efforts have saved millions of lives? No one will ever know. Many, including Bill, believed it was the usual whitewash and cover up.

* 3rd Young Eric never returned home. He didn't survive the trenches. A telegram stated he was—Killed-In-Action. One more unfortunate casualty of war, another name added to the ever-growing list of thousands killed or missing. Nothing was left of Eric but cameo memories. No one ever imagined a death toll of over sixty million. But why had so many died for so little?

* 4th In later years gonorrhoea and syphilis would be treated more sympathetically and managed with antibiotics, but not prior to 1945. It was reported during the winter of 1944 that venereal disease appeared in Bomber Command sickness reports with monotonous regularity.

* 5th Over 7,000 Avro Lancasters were built throughout the duration of the Second World War at a cost of approximately £50,000 each. About half were lost in action. Lancaster crews flew a greater number of sorties and delivered a greater number of bombs than any other aircraft. About 156,000 operations dropped in excess of 610,000 tonnes of bombs. Only thirty-five Lancasters completed more than 100 successful operations each.

  Only twenty-five crews are known to have survived their 'tour' without capture or injury.

  It was estimated about 55,000 airmen were killed. Post war debt was in the region of one billion pounds.

* 6th Bill disliked twilight until he died. He went to extreme lengths never to be alone at dusk.

* 7th The Butt Report was initially hushed up by Bomber Command. Initiated by Lord Cherwell, a personal friend of Churchill and chief scientific advisor to the cabinet.

  David Benusson-Butt was a civil servant in the War cabinet Secretariat and an assistant to Cherwell. Prepared between May 1940 and May 1941 it maintained of all bombs being dropped about half fell in open country rather than onto their designated targets. With the advent of cameras mounted under bombers triggered by the bombs release, Butt was assigned the task of assessing 633 target photographs to compare them with the crews' claims.

  The results shocked everyone. The report was devastating of its criticisms of those who were running Britain's war effort. After close examination of those aircraft recorded as attacking their target, only one in three came within a radius of five miles. Over French ports the proportion was two in three. Over Germany as a whole, the proportion was one in four. Over the Ruhr it was only one in ten. During the full moon, proportion was two in five. When there was a new moon it was as low as one in fifteen. The report claimed that all of these statistics related only to aircraft recorded as actually attacking the target.

The RAF hierarchy argued the Butt Report was inaccurate and commissioned their own report in contradiction. As a result they were allowed to keep their allocation of war materials and extended budgets.

* 8th Bill passed his seven shot Ruger .22 calibre revolver to his only son, the author, in 1965.

* 9th Bill referred often to the unfortunate friendly his crew shot down. It was an issue he regretted deeply. He never found out who it was? Might it have been Duncan? He was angry with himself that he lacked the courage after the war to pursue it and try to find out.

* 10th The Germans sent about 12,000 of these unmanned flying bombs aka Doodlebugs over South East England to principally target London.

* 11th Bill's 'logbook' was supposed to have been handed in after the cessation of hostilities. Instead he kept it. A claim was made that it had been donated to the Heritage Centre at RAAF Amberley, Australia by the executor of Bill's estate after his death in 1996. A recent search failed to recover the logbook. And the target words; 'women and children?'

The author is confident his father wrote those words.

* 12th Bill flew as a passenger often. Interestingly, he never experienced any headaches, never blacked out. He fared better health wise than most other passengers. Because he didn't like using the little bathroom on an aeroplane he never flew far.

'Fear of feeling like a busted flush, Son,' Fred had said.

Bill never suffered from de-compression or any other problems associated with flying high. He knew the Establishment at work was a powerful machine and a destructive force. The military proved to be the ultimate in that regard. He also realised he'd never worked the system in a way to gain promotion.

* 13th True to his word Bill never accepted nor claimed his campaign medals.

In December 2009 as part of the author's research, Bill's three medals were issued posthumously.

1939-45 STAR

FRANCE & STAR

WAR MEDAL 1939-45

Lest we forget

* 14th The bombing of Dresden earned the RAF the dubious distinction of the greatest area ever destroyed in a single night during the war. The city had been completely unprotected allowing the RAF to drop more than 2,600 tons of bombs to utterly destroy thirteen square miles of the historic centre.

Many years later in Brisbane, Australia, the author talked with the son of a survivor from Dresden. He painted a horrific picture of what his mum had lived through, and had subsequently told him. Heat of the burning city was so intense the air became a maelstrom of burning chunks of wood, metal and paper. Blazing tree limbs flew past heads of survivors as they fled. All kinds of objects were caught in the searing tornado, including helpless people. Tarmac roads melted. People were trapped and left to burn and suffocate. Streets were like ovens. People screamed as they tried to flee through the viscous quagmire. Shoes stuck in molten tar. Bare feet burned so badly they could no longer move.

Those citizens who were trapped died, the lucky ones, quickly.

He estimated casualties ranged between 100,000 and 160,000 people. Many died of simple asphyxiation as the firestorm sucked oxygen out of the air in overcrowded basement shelters. It took over six hours for the firestorm to begin to die down after the last raid.

By midnight fires burned so extensively the blaze was too much for the city's force of 1,000 fire fighters. Even though they were well equipped they were completely powerless. Those able, by almost some animal instinct, headed for the river. There they tried to hide under park benches and behind trees from strafing and machine guns only eighty-four days before the end of the Second World War.

About the Author

John Hickman was born in 1945 London, England. Carole, his wife of forty-five years says, 'that should make him a mature writer.'

Educated at private schools he was editor of his school magazine. Unable to secure a University scholarship he joined Westminster Bank, and then became an hotelier in Royal Tunbridge Wells.

John joined the Pest Control industry based in the UK before migrating with his family in 1971 as 'ten pound Poms.' They became naturalised in 1973.

After retirement in 2003 he started to 'smell the flowers' and realised how most of them smelled the same. Unable to play golf, he discovered a passion for writing.

Reluctant Hero is a true story about John's dad, Bill, who was a Lancaster bomber pilot in the Second World War.

As author I would like to thank you for taking the time to read my story *Reluctant Hero.*

If you enjoyed it please tell your friends.

Watch out for my next book *Sex, Knives & Crazy People* to be published late 2012.

You can connect with me online through my blog at:
www.authorjohnhickman.com

My e-mail address is: AuthorJohnHickman@gmail.com

You can also follow me on 'Twitter' at: WriterJohnH

Or join me on Facebook: AuthorJohnHickman@facebook.com

Made in the USA
Monee, IL
13 September 2026

62981010R00193